Practical Management of Chemicals and Hazardous Wastes

An Environmental and Safety Professional's Guide

W. Lee Kuhre

Seagate Technology

Senior Director Environmental, Health, Safety and Security

University of San Francisco

Graduate Program Senior Lecturer

Prentice Hall PTR
Englewood Cliffs, New Jersey 07632

Library of Congress Cataloging-in-Publication Data

Kuhre, W. Lee. 1947–
    Practical Management of chemicals and hazardous wastes : an environmental and safety
    professional's guide / W. Lee Kuhre.

       p. cm.

    Includes bibliographical references and index.

    ISBN 0-13-103946-6

    1. Hazardous wastes—Management.   2. Hazardous substances—Management.   I. Title.

TD1030.K85   1994   94-8429

604.7—dc20   CIP

Editorial/Production Supervision: Lisa Iarkowski
Interior Design: Lisa Iarkowski
Acquisitions Editor: Mike Hays
Manufacturing Manager: Alexis R. Heydt
Cover Design: Wanda Lubelska

© 1995 Prentice Hall PTR
Prentice-Hall, Inc.
A Paramount Communications Company
Englewood Cliffs, NJ 07632

The publisher offers discounts on this book when ordered in bulk quantities.
For more information, contact:

    Corporate Sales Department
    PTR Prentice Hall
    113 Sylvan Avenue
    Englewood Cliffs, NJ 07632
    Phone: 201-592-2863
    FAX: 201-592-2249.

Printed in the United States of America

10 9 8 7 6 5 4 3 2 1

ISBN 0-13-103946-6

Prentice-Hall International (UK) Limited, London
Prentice-Hall of Australia Pty. Limited, Sydney
Prentice-Hall of Canada, Inc., Toronto
Prentice-Hall Hispanoamericana S.A., Mexico
Prentice-Hall of India Private Limited, New Delhi
Prentice-Hall of Japan, Inc., Tokyo
Simon & Schuster Asia Pte. Ltd., Singapore
Editora Prentice-Hall do Brasil, Ltda., Rio de Janeiro

To Marjorie Allan Kuhre,
who encouraged and supported more than anyone.

CONTENTS

LIST OF FIGURES

LIST OF TABLES

PREFACE

Hazardous material and waste management is a topic that touches all lives. Every day people encounter hazardous material (chemicals) or hazardous waste in their home, school or place of business. How one manages those inevitable encounters is the topic of this book. Careful management of hazardous material and waste is of paramount importance if life forms are to continue to exist.

This book was written to help the environmental and safety student learn about the field and to help the working professional manage hazardous material and waste issues. For example, one issue that will impact virtually all of these people mentioned is the upcoming environmental standardization movement. The International Standards Organization (ISO) is in the process of adding comprehensive environmental and hazardous waste management systems to their future certification requirements. Most industries worldwide will be working hard to achieve this new level of environmental management. This book presents many of the systems needed to receive certification.

In order to properly manage hazardous waste, it is important to consider the entire life cycle, including when the waste was a useful chemical or hazardous material. Waste minimization is built upon this concept. Understanding the entire life cycle is also important in terms of liability, since many regulations hold generators responsible from "cradle to grave." This book takes the life-cycle concept even further, in order to provide additional insight. The discussion starts with the "conception" of the chemical and traces its evolution into a waste and even past disposal. At this point the story continues into the "afterlife," where responsibility still remains.

There is an incredible amount of important detail in this subject that covers many fields of study, including biology, chemistry, engineering, geology and other disciplines. This book was written to provide a framework in which the reader can easily integrate the detail from other references. To facilitate this integration, the second half of the book is organized around the "conception to afterlife" journey of a hazardous substance.

The author is indebted to Jane C. Kuhre and the rest of his family for their support; to Dr. Joseph Petulla of the University of San Francisco for suggesting that the book be published; to Robb Kundtz of Seagate Technology for his encouragement; to Mary Haber, Attorney at Law, for providing legal review; to George Rodgers, Eve Levin, Kent Steele, David Wilkins, Walter

Leclerc, Neil Riley, Kevin Bricknell, Eileen Bullen, Christina Scott, Sandy Stevens, Balbir Gosal, Matthew Nushwat, Christine Ryan and numerous other University of San Francisco M.S. students for providing technical suggestions; and to Michael Hays of Prentice Hall for agreeing to publish this book.

It would be physically impossible for this book to identify and address all applicable regulations for all countries. Therefore, the reader should go beyond the general management suggestions in this book and consider the relevant regulations for their specific industry and area. Even with this is mind, the author has tried to be as accurate and complete as possible. Since the subject is broad, there may be some errors and omissions, for which the author, publisher, University of San Francisco and Seagate Technology cannot be held responsible.

FOREWORD

An "environmentalist" company is one that strives to move beyond government compliance and ahead of its industry. Challenged by the need to protect and preserve our planet, consumers are changing the ways they work, play and live, and industries are changing the way they do business. Environmental management adds a critical new element to the requirements of total corporate responsibility. And, as environmental concerns become a greater part of the social conscience, environmental performance is becoming a key influence on the attitudes of shareholders.

Whether driven by the carrot or the stick, businesses, their customers and investors will recognize the potential financial impact of environmental performance, policies and actions. They will see that in the future, environmental deeds will speak louder than environmental words. Business must have a policy reflected in specific accomplishments related to the letter and spirit of the law. There are both civil and criminal penalties associated with the failure to meet environmental regulations and standards.

Given the specific challenges associated with our quest for a cleaner environment in an era of technologic growth, this is an issue that will continue to draw focus among environmental advocates, consumers and investors. Environmentally conscious companies don't wait to react to government compliance regulations that are meant to protect our habitat. Companies that strive to move beyond the mandated requirements and make efforts to prevent pollution and reduce waste cannot only make a visible difference in our environment, but in their future profitability.

Robert A. Kundtz
Senior Vice President, Administration
SEAGATE TECHNOLOGY, INC.
Scotts Valley, California 95066

CHAPTER

ONE

Introduction

Objectives

Picture yourself spending several million dollars of your organization's money cleaning up someone else's hazardous waste. Does this make you angry or nervous? Or how about being held responsible for a death or serious injury because of improper hazardous material or waste handling? Does the thought of being the defendant in a lawsuit or a prison sentence for polluting the environment make you worry? All of these scenarios are still occurring more than you would like to know. Even if they weren't, the majority of the population sincerely wants to protect the environment and the health and safety of workers.

As these words are being written, there is a monumental worldwide movement of great importance occurring. It is a new wave of environmental standardization, which will soon have a tremendously positive impact on the environment and employee health and safety. At its core will be proactive hazardous material and waste management, light-years ahead of present systems. It will be so far above the regulations that the negative incentives of hazardous material and waste management mentioned above, such as prison sentences, will pale in comparison.

This environmental standardization is just beginning to grow out of the International Standards Organization (ISO) 9000 quality certification movement. Most companies worldwide are being required by their customers to be ISO 9000 certified concerning product quality. By 1995 or 1996, this certification is expected to extend into every aspect of environmental protection.

1

Two models being considered for building this new ISO certification are British Standard (BS) 7750 and American National Standards Institute (ANSI) guidelines. When this happens, industry will be protecting the environment to a much greater degree and doing it because customers demand it as well as the regulators.

This book was written to give the reader some of the tools necessary to meet the challenges of this new environmental standardization movement and to minimize some of the risks mentioned above. Potential liability can never be completely eliminated, no matter how much analysis is done. Problems manifest themselves in an infinite number of ways and time frames. As regulations, court cases and environmental strategies become increasingly complex, it is to everyone's benefit to approach these challenges with a basic knowledge of hazardous material and waste management.

Hazardous material and waste management is an extremely broad topic. This book summarizes the essence of many scientific fields that pertain to hazardous material and waste and therefore provides a framework which will help the professional and the student organize the multitude of specifics they will encounter over the years. The various subject areas or fields important in hazardous material and waste (HMW) management include environmental sciences, engineering, industrial hygiene, law, toxicology, chemistry, biology, geology, epidemiology, preventative medicine and public relations.

In addition to federal, some state and local regulations, California regulations were included in this book as an example of state legislation. Even if readers work in other states or countries, the California regulations should still be of interest since they act as an example and often provide a basis for the legislation of many other areas.

Document Organization

Chapters 1 through 8 present some topics needed to build a basic understanding of HMW management. Chapters 9 through 12 follow the flow of a typical hazardous material/waste. Chapters 13 through 17 bring in some general environmental management ideas that help to control or add dimension to the entire process.

The case studies presented at the end of most chapters illustrate some of the topics covered in the chapters through real-life occurrences. Many of the case studies actually touch issues addressed in several chapters.

Evolution of Hazardous Material and Waste Management

Hazardous material and waste management can be divided into four stages in terms of evolution. The first stage was the realization of environmental problems. The next stage was the birth of numerous regulations designed to limit discharges to the environment, impose chemical controls and initiate investigation of past problems. The third phase emphasizes the development of waste minimization, and the fourth phase is environmental standardization.

The Realization Phase

There was a growing awareness of environmental problems for many years, but it wasn't until the 1960s when the first real period of hazardous waste awareness crystallized. The 1960s ushered in a new generation that questioned many values, including the lack of environmental concern. In 1962, Rachel Carson's *Silent Spring*, sharpened the new environmental awareness and graphically pointed out that man is damaging the environment with hazardous materials and wastes. Many people really started to become aware that some of the past and present practices concerning hazardous waste might be causing significant damage. For example, midnight dumping and accidental spills were becoming known. The use of unlined pits, ponds and lagoons and the use of certain hazardous waste disposal sites were starting to be questioned. Discharges of untreated waste into sewer systems and streams were causing people to ask embarrassing questions.

The Regulation Phase

A second and unique period of time extended from 1969 through 1980, when significant new environmental legislation was introduced to address some of the concerns that surfaced most loudly in the 1960s. For example, the National Environmental Policy Act (1969), major Clean Air Act amendments in 1970 and 1977, the Resource Recovery Act (1970), the Poison Prevention Packaging Act (1970), the Consumer Product Safety Act (1972), the Federal Water Pollution Control Act (1972), the Federal Insecticide, Fungicide and Rodenticide Act Amendment (1972), the Marine Protection Research and Sanctuaries Act (1972), the Safe Drinking Water Act (1974), the Hazardous Materials Transportation Act (1975), the Resource Conservation and Recovery Act (1976), the Toxic Substance Control Act

(1976), the Uranium Mill Tailings Radiation Control Act (1978) and the Comprehensive Environmental Response, Compensation and Liability Act (1980) were all enacted. Never in history, even up to the present day, have so many far-reaching environmental acts been introduced and in a period of only 11 years. Other legislation has been enacted since, but it pales in comparison to these monumental acts. This period of time forever changed the course of history for hazardous material and waste management. Also during this period of time, the importance of the acts mentioned above was underlined when contamination became obvious at Love Canal in 1978, the Hardeman County Landfill (Tennessee) in 1972 and the La Bounty Dump (Iowa) in 1977.

The Waste Minimization Phase

The beginning of the waste minimization period started years ago in Europe and the Far East, but it did not get a great amount of attention in the United States until around the mid-1980s. Events such as Bhopal in 1986 emphasized many lessons, including the minimization of hazardous material and waste. The Superfund Amendments and Reauthorization Act (1986) required people to inventory and disclose their hazardous materials; and, to their horror, they had much more than imagined [Bricknell 1990]. One obvious solution was to minimize the hazardous materials and wastes. Many organizations voluntarily made great progress with minimization in the 1980s and 1990s. Others basically offered only promises. California has adopted a hazardous waste minimization program (Hazardous Waste Source Reduction and Management Review Act of 1989, Health and Safety Code 25244.12-25244.24). A large amount of sincere environmental interest and initiative is still needed, however, since the waste minimization requirements are still considered weak in comparison to other hazardous material and waste regulations.

Environmental Standardization Phase

The fourth phase of HMW evolution is the present environmental standardization phase. This started in earnest in the early 1990s with ISO 9000 certification. ISO 9000 certification is basically voluntary certification of product quality by outside auditors. Environmental, health and safety are excluded; however, starting around 1995 or 1996, this will all change dramatically. Various environmental guidelines are being considered from different parts of the world for inclusion into ISO. For example, the American National Standards Institute guidelines and British Environmental Management Standard (BS) 7750 may be used and would make proactive and extremely comprehensive environmental, health and safety programs a requirement of future ISO 9000 certification. Overall, this is going to move environmental protection, health and safety light-years ahead in a very short time frame.

The new ISO environmental standards will probably require four massive documents. The first document would be named the Register of Regulations and include local, regional, state, federal and international regulations that apply to each operation. Chapter 2 of this book provides much of the information needed for certification. The second document may be the Register of Effects, which is a collection of all the issues and impacts created by the operation that impact the environment. Hazardous material and waste, the topic of this book, create the majority of impacts for most operations. The third document would be the Operations Procedures Manual, which is a collection of the management systems to address the regulations and impacts and is the primary objective of this book. The last document probably will be the Environmental Management Manual, which presents policy statements, vendor management, auditing and other miscellaneous systems. Again, these topics are detailed in this book.

General Types and Sources of Hazardous Waste

Table 1-1 presents an overview of where most RCRA designated hazardous waste can be found. For example, in 1985 Texas led the nation with 38,006,000 tons generated, followed by Georgia, Tennessee and Pennsylvania. What Table 1-1 does not show is non-RCRA hazardous waste, such as California-regulated hazardous waste, which is considerable. The source of the data is from 1985 Biennial Reports and from records provided by treatment, storage and disposal facilities. Table 1-1 is useful for indicating rough estimations of hazardous waste volumes in each state for comparison purposes; however, many states define hazardous waste in slightly different ways.

Figure 1-1 breaks RCRA hazardous wastes generated in 1986 into five categories. These include—from greatest volume to least—inorganic liquids, inorganic sludges, organic liquids, organic sludges and inorganic solids. The volumes shown in the figure are in million metric tons (MMTs) generated per year. The fact that the volume of inorganic liquid hazardous waste is so much larger (252 MMT) than the other types (2–15 MMT) has

Table 1.1   State-by-State Profile of RCRA Hazardous Waste Generated
Exported and Imported
(Figures in thousands of tons)

State	Quantity of RCRA Waste Generated	Shipped Out of State (Exported)	Received (Imported)
Alabama	7,403	66	146
Alaska	3	1	
Arizona	66	16	3
Arkansas	57	53	29
California	3,385	4	33
Colorado	295	22	1
Connecticut	190	76	37
Delaware		70	9
District of Columbia	2	2	
Florida	834	104	12
Foreign			
Georgia	37,325	76	53
Guam	>1		
Hawaii	7	>1	
Idaho	2	2	8
Illinois	2,141	118	676
Indiana	2,518	695	162
Iowa	121		5
Kansas	1,325	11	13
Kentucky	7,600	55	55
Louisiana	12,182	103	368
Maine	7	7	
Maryland	698	103	90
Massachusetts	114	157	17
Michigan	4,077		267

Table 1.1 State-by-State Profile of RCRA Hazardous Waste Generated
Exported and Imported (Continued)
(Figures in thousands of tons)

State	Quantity of RCRA Waste Generated	Shipped Out of State (Exported)	Received (Imported)
Minnesota	61	30	
Mississippi	2,507	83	25
Missouri	64	44	10
Montana	25	>1	
Nebraska	543	14	1
Nevada		1	2
New Hampshire	20	12	14
New Jersey	8,653	311	152
New Mexico	9	2	3
New York	444	138	187
North Carolina	1,285		25
North Dakota	3	3	2
Ohio	2,966	263	340
Oklahoma	1,591		56
Oregon	31	9	64
Pennsylvania	31,307	261	383
Puerto Rico	149		
Rhode Island		9	32
South Carolina	5,033	18	120
South Dakota	1	>1	>1
Tennessee	33,199	53	24
Texas	38,006	197	99
Utah	1,135		9
Vermont	10	11	>1
Virginia	24,994	108	19

Table 1.1 State-by-State Profile of RCRA Hazardous Waste Generated
Exported and Imported (Continued)
(Figures in thousands of tons)

State	Quantity of RCRA Waste Generated	Shipped Out of State (Exported)	Received (Imported)
Washington	338		9
West Virginia	12,077	62	18
Wisconsin	85	40	20
Wyoming	16	>1	1
Total	**245,128**	**3,732**	**3,627**

Source: Draft 1985 National Biennial Report of Hazardous Waste Generators and Treatment, Storage and Disposal Facilities Regulated under RCRA (January 1988). Reprinted from EPA *The Waste System,* Washington, 1988.

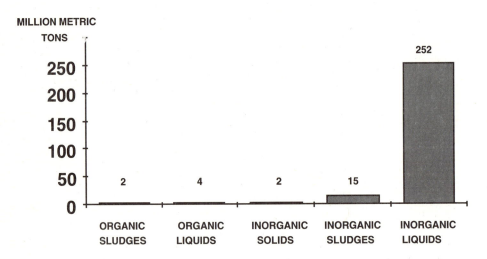

Figure 1.1 Estimate of Physical Characteristics of RCRA Hazardous Wastes. (From EPA, *The Waste System,* Washington, 1988, National Screening Survey.)

resulted in the development of a huge industry to handle this type of waste. For example, there are numerous filtration equipment manufacturers that help generators remove as much of the liquid as possible to produce a smaller volume in the form of inorganic sludges.

Highlights of Hazardous Material and Waste (HMW) Management

There are literally hundreds of specific things that should be done to properly manage HMW. First, there must be a serious commitment from senior management to HMW management. This should be clearly shown with a written policy statement and by providing the resources necessary. Second, the awareness of all employees must be raised and kept up with continual reminders. One way many organizations accomplish this is by providing employee reminders, possibly an environmental video featuring top-level management. Third, the regulations need to be boiled down into a meaningful set of procedures and written in lay terms specific to the operation. Employees must be familiar with the procedures and follow them. Fourth, employees must be trained concerning HMW management. Having a procedure book is not enough; they need classroom and on-the-job training to make it work. An audit system must be implemented to identify potential HMW problems before they become unmanageable. Highly efficient technology and equipment should be used and maintained. These key elements are discussed in greater detail in various chapters of this book.

Some Basic Definitions

A few basic acronyms with working definitions are presented here, which are key for understanding the concepts in the rest of this book. A list of additional acronyms, arranged in alphabetical order for easy reference, can be found on page 320. Additional definitions can be found in the glossary at the end of this book. Working definitions are presented for clarity. The actual legal definitions can be found in the regulations. In a few cases, the working definition practiced in real life does not equal the legal definition.

Hazardous Materials (HM)   Chemicals that are capable of posing an unreasonable risk to health, property or the environment when stored, used or transported in commerce. They are substances that have or will have functional uses. Some examples are gases, chlorine, fertilizers, fuels, acids, solvents, and cleaning compounds. Hazardous materials are commonly defined by agencies as substances in a quantity or form posing an unreasonable risk to health, safety and/or property when stored, used or transported in commerce.

Hazardous Wastes (HW)   Waste substances that by nature are inherently dangerous to handle or to dispose and may cause death, illness or be a hazard to health or the environment when improperly handled.

These are hazardous materials that have been used, spilled or are no longer needed. Examples are old explosives and waste products from chemical processes.

Many agencies have different definitions of hazardous waste. Probably the most universally accepted definition is in RCRA and includes waste or combinations of wastes, which, because of its quantity, physical, chemical or infectious characteristics may (1) cause or significantly contribute to an increase in mortality or an increase in serious irreversible or incapacitating reversible illness, or (2) pose a substantial present or potential hazard to human health or the environment when improperly treated, stored, transported or disposed.

Hazardous Material and Hazardous Waste (HMW)   HMW refers to a chemical (hazardous material) and a *hazardous* waste. Please take special note of this abbreviation. HMW is used more than any other acronym throughout this book.

Toxic Waste   A waste that is capable of killing, injuring or damaging an organism. Many incorrectly use the term interchangeably with hazardous waste. Toxic waste is a type of hazardous waste. It is also any pollutant listed as toxic under Section 307(a)(1) of the Clean Water Act.

Generator   Any person or organization whose act first causes a hazardous material to become a hazardous waste subject to regulation. An organization usually becomes a generator when its process generates a waste or when it spills or no longer uses a hazardous material.

Transporter   The carrier of a hazardous material or waste. The carrier can also be a generator if the carrier's vehicle spills, crosses country boundaries or accumulates sludges or residues of the materials transported.

An International Overview of HMW Management

HMW management is as different as night and day from one part of the world to the next. Throughout this book, specific examples will be given to show some of the differences that exist between the U.S., Europe and the Far East. It is hard to generalize whole countries, however, so some rough observations will be stated for Europe based on the author's work in Scotland, Portugal, the Netherlands, Ireland and Germany. The generalizations made for the Far East will be based on personal experiences in Indonesia, Singapore, Malaysia and Thailand. The U.S. statements will be based primarily on the author's experience in California, Colorado, Utah, Nevada, Minnesota, Nebraska, Texas and Oklahoma.

Table 1.2 An International Comparison
of Hazardous Material and Waste Management

	Far East	Europe	United States
Practical Action (Pragmatic)	Low	High	Moderate
Enforcement Action	Low	Moderate	High
Litigation Driven Action	Low	Low	High
Number of Applicable Regulations	Moderate	Moderate	High
Pollution Level	High	Moderate	Low

The above is based upon observations made by the author in the following countries or states:
Far East = Singapore, Thailand, Malaysia, Indonesia
Europe = Portugal, Scotland, the Netherlands, France, England, Germany, Ireland
U.S. = California, Oregon, Utah, Colorado, Ohio, Minnesota, Nebraska, Montana, Oklahoma

Table 1-2 presents a few initial and very general differences in HMW management in these three parts of the world. Obviously, there are many exceptions to what the table shows. Of greatest importance is the last row of the table, pollution level. From the author's observations, pollution from HMW seems to be less in the U.S., greater in Europe and greatest in the Far East. This may correlate directly with the approximate number of HMW regulations shown in the fourth row. Overall, the U.S. generally seems to take care of HMW because of regulations and litigation. Europe manages HMW because it is a practical thing due to a critical shortage of space and resources. Certain parts of the Far East, in general, have only started to deal with the HMW it creates.

Life Cycle of an HMW

Figure 1-2 presents the life cycle of an HMW from conception through afterlife. The most important aspect of the figure is the evolution of the hazardous material into a hazardous waste. This illustration provides a framework that ties together many of the concepts in this document. The life cycle, or ordering of events, is also used to organize and present the regulations in Chapter 2 and most of the chapters in the second half of the book.

At the bottom of Figure 1-2 is an arrow, entitled "Treated discharge to the air." This number was estimated to be 1.8 billion pounds of toxic chemicals discharged to the air in 1987 from point sources. Next to that

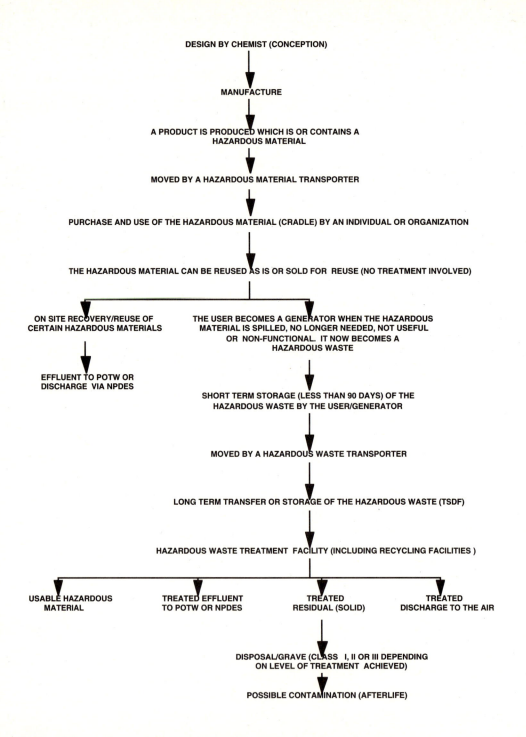

Figure 1.2 Life Cycle of a Hazardous Material/Waste.

arrow is one entitled "Treated effluent to POTW (sewer/Public Owned Treatment Works) or NPDES (stream/National Pollutant Discharge Elimination System)." The amount that went to POTW was estimated to be 884 million pounds in 1987 and to surface waters via NPDES, 555 million pounds. These figures are for manufacturing facilities only and indicate the general magnitude of the discharges [Blackman 1993].

The Importance Of Understanding Basic Chemistry

A basic understanding of chemistry is essential for effective HMW management. Chemical properties and reactions of both the inorganic and organic types are part of the thought process at almost every step of the HMW life cycle. Obviously, the creator and manufacturer utilize chemical principles when the hazardous material is first introduced. The transporter and emergency response (ER) teams must consider the chemistry of the hazardous material, especially if it is spilled. Proper storage and use of hazardous materials requires a basic understanding of chemical reactions. Last, treatment and disposal scenarios all heavily involve chemical properties and reactions.

Key References

HMW references are numerous. The reference list at the end of this book presents a few of these sources of information, which were utilized in the writing of this book. Four common references of special importance are the Federal Register (FR), the Code of Federal Regulations (CFR), state hazardous waste regulations such as the California Code of Regulations (CCR), and Material Safety Data Sheets. Table 1-3 lists additional common references.

Table 1.3   Sample of Sources of Hazardous
Material/Waste Information

- *Federal Register.*
- *Code of Federal Regulations.*
- *Hazardous Materials Transportation Guide.*
- *Deering's California Codes, Health and Safety,* Current.
- *Hazardous Communication Handbook,* California Chamber of Commerce, *1986.*
- *Hazardous Waste Management Handbook, Guidelines for Compliance,* California Chamber of Commerce, 1987.

Table 1.3 Sample of Sources of Hazardous
Material/Waste Information (Continued)

■ *Industrial and Hazardous Waste Management Firms,* Environmental Information LTD, 1986.

■ *Hazardous Waste Audit Program,* J.J. Keller, 1982.

■ *Chemical Emergency Preparedness Program,* EPA, 1985.

■ *Dangerous Properties of Industrial Materials,* Sax.

■ *Hazardous Materials, Substance and Wastes Compliance Guide,* EPA, 1984.

■ *Hazardous Waste Regulation Handbook,* Weinberg, 1983.

■ *Hazardous Waste Services Directory,* Keller, 1983.

■ *Registry of Toxic Effects of Chemical Substances,* NIOSH, 1980.

■ *General Industry Safety Orders.*

■ *California Code of Regulations Title 22, 8, 23, etc.*

■ *Toxics News,* Capitol Reports, 1988.

■ *California OSHA Reporter,* Sten-o-Press, 1988.

■ *OSHA Compliance Advisor,* Business and Legal Reports

■ *California's Proposition 65,* California Chamber of Commerce, 1988.

■ *Fire Protection Guide on Hazardous Materials,* 9th ed., National Fire Protection Association, 1986.

■ *Hawley's Condensed Chemical Dictionary,* 11th ed., Sax and Lewis, 1987.

■ *Hazardous Chemicals Desk Reference,* Sax and Lewis, 1987.

The Federal Register

The Federal Register is a chronological publication of proposed and final federal regulations pertaining to almost all subjects, including the environment. It also contains executive orders, notices of hearings, acceptances of responses and interpretations. An example of the cover of a Federal Register is shown in Figure 1-3. Review of the Federal Register is a very tedious process because of its volume and coverage of so many subjects. There are many review services on the market today that extract and summarize various Federal Register subjects, such as hazardous waste.

Code of Federal Regulations (CFR)

The CFR is a subject publication of final federal regulations. It is divided into 50 titles. Each title is published once a year and can be obtained

Figure 1.3 *Federal Register.*

from the Government Printing Office. The CFRs are kept up to date via the Federal Register. A subject index to the CFR is contained in a separate volume. CFRs of special importance include 49 CFR (transportation of HMW), 40 CFR (protection of the environment) and 29 CFR (OSHA, or employee safety). An example of a CFR cover is shown in Figure 1-4.

State or Regional Hazardous Waste Regulations

State or regional regulations are very important in the management of HMW. For example, the California Code of Regulations (CCR) covers many aspects of California's legislation, including environmental. It is very similar to the CFR in organization. Titles of special HMW interest include 22, 23 and 26. The CCR has been rewritten to conform with RCRA. Title 26 summarizes many of the topics presented in Titles 22 and 23.

Material Safety Data Sheets (MSDS)

The HMW manager should request from the manufacturer an MSDS for every chemical purchased or proposed for purchase. These MSDSs must then be made available upon request to any employee in the organization.

Section 1910.1200 of 29 CFR requires all chemical manufacturers or importers to assess the hazards of chemicals that they produce or import. These regulations are enforced by the Occupational Safety and Health Administration (OSHA). In addition to OSHA, other agencies depend on the MSDS as a source of information for employees and the community. The primary purpose of the MSDS is to protect employees through the provision of exposure, handling, storage and disposal information. Many MSDSs have just recently begun to address environmental issues. An example of an MSDS is shown in Figure 1-5.

One of the most useful parts of an MSDS for the HMW manager is the National Fire Protection Association (NFPA) hazard identification. This can be seen in the upper right portion of Figure 1-5. A person can quickly see if there is a rating of 3 or 4 in terms of health, fire or reactivity hazard. If so, the chemical should be handled with care. On the other hand. if it has a rating in the 0–2 range, there is not as much concern.

There are several other important entries on the MSDS that tie closely together. The percent composition of the constituents along with the Chemical Abstract Service (CAS) numbers are related and very important. The CAS number can be checked in other chemical books to get more details about each constituent. Without the CAS number, it is often hard to

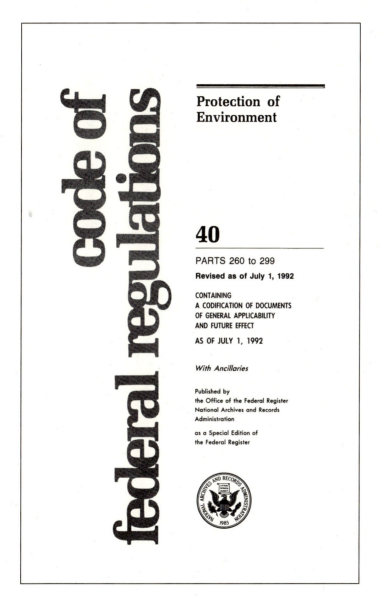

Figure 1.4 *Code of Federal Regulations.*

find the precise chemical in the reference document. Permissible Exposure Limits (PELs) and Threshold Limit Values (TLVs) should be on the MSDS for each constituent. This data provides information about the toxicity of

MATERIAL SAFETY DATA SHEET

PENNZOIL COMPANY

I PRODUCT IDENTIFICATION

Manufacturer's Name	PENNZOIL COMPANY	CAS Number: MIXTURE
		MSDS Code: 000950

Address	P.O. BOX 808 OIL CITY, PENNSYLVANIA 163010808

NFPA Hazard Identification

Degree of Hazard	Hazard Ratings
Health: 1 Fire: 1 Reactivity: 0	0 - Least 1 - Slight 2 - Moderate 3 - High 4 - Extreme

Emergency Telephone No.	(713) 236-6070
Trade Name	PENNZOIL ANTI-FREEZE & SUMMER COOLANT
Synonyms	ANTI-FREEZE

II INGREDIENTS

COMPONENT NAME CAS NUMBER	HAZARDOUS IN BLEND	PERCENTAGE MIN MAX	COMPONENT EXPOSURE LIMITS	UNITS
ETHYLENE GLYCOL 107-21-1	YES	80.00 TO 95.00	ACGIH TLV 125.0000	MG/M3
WATER 7732-18-5	NO	1.00 TO 5.00	OSHA PEL ACGIH TLV	NO LIMIT NO LIMIT
ADDITIVE PACKAGE CONTAINING SODIUM PROPRIETARY	NO	1.00 TO 5.00	OSHA PEL ACGIH TLV	NO LIMIT NO LIMIT
ADDITIVE PACKAGE CONTAINING PHOSPHOROUS PROPRIETARY	NO	1.00 TO 5.00	OSHA PEL ACGIH TLV	NO LIMIT NO LIMIT
ADDITIVE PACKAGE CONTAINING SODIUM, SILICON PROPRIETARY	NO	< 1.00	OSHA PEL ACGIH TLV	NO LIMIT NO LIMIT
ADDITIVE PACKAGE CONTAINING SODIUM, NITROGEN PROPRIETARY	NO	< 1.00	OSHA PEL ACGIH TLV	NO LIMIT NO LIMIT

BY: ENVIRONMENTAL, SAFETY & HEALTH (713) 546-8512

EFFECTIVE DATE: NOVEMBER 15, 1985

DISCLAIMER OF WARRANTY:

THE INFORMATION CONTAINED HEREIN IS BASED UPON DATA AVAILABLE TO US, AND REFLECTS OUR BEST PROFESSIONAL JUDGEMENT. HOWEVER, NO WARRANTY OF MERCHANTABILITY, FITNESS FOR ANY USE, OR ANY OTHER WARRANTY IS EXPRESSED OR IMPLIED REGARDING THE ACCURACY OF SUCH DATA, THE RESULTS TO BE OBTAINED FROM THE USE THEREOF, OR THAT ANY SUCH USE DOES NOT INFRINGE ANY PATENT. SINCE THE INFORMATION CONTAINED HEREIN MAY BE APPLIED UNDER CONDITIONS OF USE BEYOND OUR CONTROL AND WITH WHICH WE MAY BE UNFAMILIAR, WE DO NOT ASSUME ANY RESPONSIBILITY FOR THE RESULTS OF SUCH APPLICATION. THIS INFORMATION IS FURNISHED UPON THE CONDITION THAT THE PERSON RECEIVING IT SHALL MAKE HIS OWN DETERMINATION OF THE SUITABILITY OF THE MATERIAL FOR HIS PARTICULAR PURPOSE.

REQUIRED UNDER USDL SAFETY AND HEALTH REGULATIONS FOR SHIP REPAIRING, SHIPBUILDING, AND SHIPBREAKING (29 CFR 1915, 1916, 1917).

Figure 1.5 Material Safety Data Sheet.

the chemical. Other items to note on the MSDS include physical characteristics, primary routes of entry, precautions, controls, emergency and first aid procedures and the name and phone number of the manufacturer.

Other Sources of Information

Hazardous Waste Haulers, Recyclers and Certified Labs

Various lists for haulers, recyclers, disposal sites and labs are available from the EPA or the lead state HMW agency. In addition to the list mentioned above, the lead state agency also has lists of Registered Environmental Assessors, Waste Exchanges, Chemical Lists and so on.

Important Phone Numbers

When an emergency or question arises, it is important to quickly know who to call. Table 1-4, Federal Hazardous Material and Waste Information Numbers, is a detailed list of many useful phone numbers. Here are a few of the most commonly used federal phone numbers for HMW questions:

RCRA/Superfund Hotline: 800-424-9346

National Response Center: 800-424-8802

Chemical Transportation Emergency Center:
    800-424-9300

Community Right-To-Know (SARA): 800-535-0202

Table 1.4   Federal Information Numbers for
Hazardous Material and Waste

Chemtrec Chemical Trans. Emergency Center 800-424-9300	**National Capital Poison Control Center for Ingestion of Chemicals** 202-625-3333
National Response Center U.S. Coast Guard 800-424-8802	**National Pesticide Tele. Network** Pesticide Information Office 800-858-7378

Table 1.4 Federal Information Numbers for
Hazardous Material and Waste (Continued)

National Small Flows Clearinghouse	Office of Water
	202-382-5700
Comm. Wastewater Treatment	
Information	Pesticides Docket
800-624-8301	703-557-4434
U.S. Environmental Protection Agency	Public Information Reference Unit
	202-382-5926
Agency Locater Service	
202-382-2090	RCRA Docket
	202-475-9327
Air Docket	
202-382-7548	Superfund Docket
	202-382-3046
Center for Environmental	
Research Information	TSCA Assistance Information Office
513-569-7562	202-554-1404
Drinking Water Docket	TSCA Docket
202-475-9598	202-382-3587
FOIA Office	Asbestos Ombudsman
202-382-4048	800-368-5888
Great Lakes National Program Office	Federal Facilities Docket Hotline
312-353-2000	800-548-1016
Hazardous Waste Ombudsman	RCRA and Superfund Hotline
202-475-9361	800-424-9346
OUST Docket	SARA and R-T-K Hotline
202-475-9720	800-535-0202
Office of Air and Radiation	Safe Drinking Water Hotline
202-382-7400	800-426-4791
Office of Enforcement and Compliance	Small Business Ombudsman
202-382-4134	800-368-5888
Office of General Counsel	**U.S. Environmental Protection**
202-475-8040	**Agency Library**
Office of Intergovernmental Liaison	Hazardous Waste Collection (Hdqt.)
202-382-4454	202-382-5934
Office of Pesticides and Toxics	Headquarters
202-382-2902	202-382-5921
Office of Public Affairs	National Enforcement Investigations
202-382-4361	303-236-5122
Office of Solid Waste	Research Triangle Park Library
202-382-4627	919-541-2777
Office of Superfund	
202-382-2180	

Table 1.4   Federal Information Numbers for
Hazardous Material and Waste (Continued)

**U.S. Environmental Protection
Agency Region I**

General Number
617-565-3715

Hazardous Waste Ombudsman
617-223-1461

Library
617-563-3300

R-T-K Hotline
617-565-1260

Small Business Ombudsman
617-860-4300

**U.S. Environmental Protection
Agency Region II**

General Number
212-264-2515

Hazardous Waste Ombudsman
212-264-2980

Library
212-264-2881

R-T-K Hotline
201-321-3222

RCRA Hotline
800-732-1223

Small Business Hotline
212-264-4711

Superfund Hotline
800-722-1223

**U.S. Environmental Protection
Agency Region III**

General Number
800-438-2474

Hazardous Waste Ombudsman
215-597-0982

Library
215-597-0580

R-T-K Hotline
215-597-6418

Small Business Ombudsman
215-597-9817

**U.S. Environmental Protection
Agency Region IV**

General Number (Georgia)
800-282-0239

General Number (Non-GA States)
800-241-1754

Hazardous Waste Ombudsman
404-347-3004

Library
404-347-4216

R-T-K Hotline
404-347-4727

Small Business Ombudsman
404-347-7109

**U.S. Environmental Protection
Agency Region V**

General Number (Illinois)
800-572-2512

General Number (non-IL States)
800-621-8431

Hazardous Waste Ombudsman
312-353-5821

Library
312-353-2022

R-T-K Hotline
312-886-2806

Small Business Ombudsman
312-353-2072

Table 1.4 Federal Information Numbers for
Hazardous Material and Waste (Continued)

**U.S. Environmental Protection
Agency Region VI**

Environmental Emergency Hotline
214-655-2222

General Number
214-655-2200

Hazardous Waste Ombudsman
214-655-6765

Library
214-655-6444

R-T-K Hotline
214-655-1730

Small Business Ombudsman
214-655-6580

**U.S. Environmental Protection
Agency Region VII**

Action Line (Kansas)
800-221-7749

Action Line (Non-KS States)
800-223-0425

General Number
913-236-2903

Hazardous Waste Ombudsman
913-236-2852

Library
913-236-2828

R-T-K Hotline
913-236-7054

Small Business Ombudsman
913-236-2800

**U.S. Environmental Protection
Agency Region VIII**

General Number
800-759-4372

Hazardous Waste Ombudsman
303-294-7036

Library
303-293-1444

R-T-K Hotline
303-293-1270

Small Business Ombudsman
303-294-7036

**U.S. Environmental Protection
Agency Region IX**

General Number
415-974-8076

Hazardous Waste Ombudsman
415-974-8916

Library
415-974-8082

R-T-K Hotline
415-974-3273

RCRA Hotline
415-974-7473

Small Business Ombudsman
415-974-0960

Superfund Hotline
800-231-3075

**U.S. Environmental Protection
Agency Region X**

General Number
206-442-5810

Hazardous Waste Ombudsman
206-442-2871

Library
206-442-1289

R-T-K Hotline
206-442-6765

Small Business Ombudsman
206-442-4280

Simonsen, C.B., *1994 Environmental Law Resource Guide.* This book may be purchased from the publisher, Clark Boardman Callaghan, 155 Pfingsten Road, Deerfield, Illinois 60015, Telephone 1-800-323-1336.

Case Studies

Case studies are presented at the end of each chapter. These real-life examples will help to illustrate topics covered in the chapter. Many of the case studies give examples of multiple issues addressed in several chapters. This especially applies to this first case study.

City of Yonkers

In December 1975, organized crime figures in New York signed an open-ended memorandum with the city of Yonkers to haul the City's garbage. An undercover agent with the Police Department discovered that someone at City Hall was on the take. The agent recommended several times to his boss that an investigation be made to find out where the garbage was going. The boss replied that since it was being shipped out of Yonkers, it was out of their jurisdiction.

Even though the undercover agent did not receive any support, he continued with his mission. By Spring 1976, the agent discovered that the garbage was going into an illegal landfill, which was not permitted by the Department of Environmental Conservation (DEC), and that it violated the Wetlands Act, a New York statute.

But it wasn't just garbage! On August 2, 1979, New Jersey State Police Detective Sergeant Richard Ottens photographed a worker standing on top of a mound of garbage inside of a roll-off. The worker pushed a metal drum containing holes into the center of the garbage. Next, the worker attached a two-inch hose from an 8000-gallon hazardous waste tanker and opened the release valve. Once the tanker was empty and the garbage contained its new chemical surprise, the roll-off was transported to the class III landfill. Detective Ottens knew he was photographing a crime in progress.

About three years later, a special grand jury for investigations into organized criminal activities issued a lengthy report of their findings. Seventeen private wells had been contaminated with trichloroethylene (TCE). Lab analysis was conducted on kitchen tap water and disclosed levels of TCE up to 1000 times greater than the maximum recommended concentration. A DEC official testified that the dump's chemical contents were probably contaminating the drinking water as well as the nearby Hackensack River. The DEC ordered the landfill closed.

A mafia hit man later disclosed that this class III landfill was one of his disposal grounds of choice. His victims' bodies degraded faster in it because of all the illegal hazardous waste. This minimized the chance of the bodies ever being found [Block 1985].

TWO

Agencies and Regulations

Introduction

The regulations and agencies are very important and, to a great degree, direct the show. They are initially presented here, but they are also at the heart of many other chapters. An overview of the environmental legislative process is shown in Figure 2-1. This illustration applies to environmental legislation as well as many other types of regulations. As can be seen in this figure, the federal, state and local layers are interrelated. In many cases, the whole process starts with federal legislation and works its way down. This is not always the rule, since legislation can also be introduced locally or at the state level. Additionally, in some instances a fourth layer is added when a state creates regional agencies.

One big difference that exists in HMW from one part of the world to the next is in terms of the number of regulations, agencies and enforcement. There are fewer agencies and regulations in Europe, and even fewer in the Far East, as compared to the United States. In Thailand, for example, the Enhancement and Conservation of National Environmental Quality Act, the Factories Act, the Toxic Substances Act and the Public Health Act are the primary pieces of legislation. Great Britain has the Control of Substances Hazardous to Health Regulations of 1988 and a few others.

Since the toxic regulations are relatively new in most parts of the world, there is a lack of an overall framework and many differences in interpretation of the regulations. In many cases, agencies and regulations overlap and conflict. This has resulted in a lack of coordination, fragmented agencies and conflicting

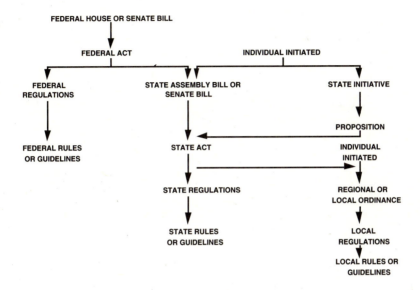

Figure 2.1 Environmental Legislative Process Overview.

answers. It may be several more years before the agencies and industry understand the existing toxic/hazardous regulations well enough to make meaningful and positive change.

Some of the regulations may never be well organized or integrated from the top down. Therefore, it is important to manage them from the bottom up. For example, if one looks for common regulations or requirements and designs programs that will comply with more than one regulation, one would be ahead of the game. For example, training is required by OSHA, RCRA and numerous state regulations. Therefore, one training program should be designed that will address all the requirements and avoid overlap.

Because of the regulatory confusion, several loopholes still exist, which generators commonly use. For example, the determination of when a hazardous material becomes a hazardous waste is open to interpretation in many cases. Recycling activities are also nonuniformly regulated.

Agencies

Interrelationship

There is considerable overlap in agency authority, responsibility and function. This has both a negative and a positive impact. On the positive side, the overlap provides a back-up or control and reduces the chance of something slipping through the cracks. On the negative side, the overlap causes duplication of effort, hard feelings between agencies and confusion.

The grand design is for the federal agencies to set the minimum standards and direction and the state, regional, county or city agencies to develop and run local, specific and usually stricter programs. It doesn't always work this way, and the federal and state agencies sometimes implement the essence of the program (Figure 2-1). Federal law has supremacy over state law in cases of conflict between federal and state, but not under shared powers. The Hazardous Waste Control Law allows states the option to regulate more strictly than federal law.

Lead Agency

Table 2-1 lists all the lead state hazardous waste agencies. Some of the names may have changed, since the governments often rearrange and rename some of their agencies. In addition, there are separate lead agencies in terms of transportation of HMW, such as the Department of Transportation (DOT) and the state highway patrol. This list becomes outdated very quickly as agencies consolidate and change their names. For example, Table 2-1 shows California's lead agency as the Department of Health Services, which it was. Now, however, the lead agency is called the Department of Toxic Substances Control under the California EPA. Recent attempts in California on Permit-Streamlining are meant to make the lead agency approach more effective.

Table 2.1   State and Territorial Hazardous Waste Management Agencies

Alabama	American Samoa	Arkansas
Land Division	Environmental Quality	Hazardous Waste Division
Alabama Department	Commission	Arkansas Department of
of Environmental Management	Government	Pollution Control and Ecology
1751 Federal Drive	of American Samoa	P.O. Box 9583
Montgomery, AL 36130	Pago Pago, American Samoa	Little Rock, AR 72219
205-271-7730	96799	501-562-7444, Ext. 504
	Overseas Operator: 663-2304	
Alaska	**Arizona**	**California**
Alaska Department of	Office of Water and Water	Toxic Substances
Environmental Conservation	Quality Management	Control Division
Division of Environmental	Arizona Department of Envi-	Department of Health Services
Quality	ronmental Quality	P.O. Box 942732
P.O. Box #O	2005 N. Central Avenue	400 P Street
Juneau, AK 99801	Room 304	Sacramento, CA 95814
907-465-2666	Phoenix, AZ 85004	916-324-1826
	602-257-2211	

Table 2.1   State and Territorial Hazardous
Waste Management Agencies (Continued)

Colorado
Waste Management Division
Colorado Department
of Health
4210 E. 11th Avenue
Denver, CO 80220
303-331-4830

**Commonwealth
of North Mariana Islands**
Division of Environmental
Quality
Department of Public Health
and Environmental Services
Commonwealth of the North-
ern Mariana Islands
Office of the Governor
Saipan, Mariana Islands 96950
Overseas Operator: 6984
Cable Address:
   Gov. NMI Saipan

Connecticut
Hazardous Material Manage-
ment Unit
Department of Environmental
Protection
State Office Building
165 Capitol Avenue
Hartford, CT 06106
203-566-4924

Delaware
Hazardous Waste Management
Section
Division of Air and Waste
Management
Department of Natural Re-
sources and Environmental
Control
P.O. Box 1401
89 Kings Highway
Dover, DE 19903
302-736-3672

District of Columbia
Pesticides and Hazardous
Materials Division
Department of Consumer and
Regulatory Affairs
5010 Overlook Avenue, S.W.
Room 114
Washington, D.C. 20032
202-783-1194

Florida
Division of Waste
Management (UST)
Department of Environmental
Regulations
Twin Towers Office Building
2600 Blair Stone Road
Tallahassee, FL 32301
904-488-0190

Georgia
Land Protection Branch
Industrial and Hazardous
Waste Management Program
Floyd Towers East/Room 1154
205 Butler Street, S.E.
Atlanta, GA 30334
404-656-2833

Guam
Hazardous Waste Management
Program
Guam Environmental
Protection Agency
P.O. Box 2999
Agana, Guam 96910
Overseas Operator:
   671-646-8863

Hawaii
Department of Health
Hazardous Waste Program
P.O. Box 3378
Honolulu, HI 96801
808-543-8226

Idaho
Hazardous Materials Bureau
Department of Health
and Welfare
Idaho State House
450 W. State Street
Boise, ID 83720
208-334-5879

Illinois
Division of Land Pollution
Control
Illinois Environmental
Protection Agency
2200 Churchill Road
Springfield, IL 62706
217-782-6760

Indiana
Indiana Department of Environmental Management
105 S. Meridian Street
P.O. Box 6015
Indianapolis, IN 46225
317-232-3210

Iowa
Air Quality and Solid Waste
Protection
Department of Water, Air, and
Waste Management
900 East Grand Avenue
Henry A. Wallace Building
Des Moines, IA 50319-0034

Kansas
Bureau of Waste Management
Department of Health
and Environment
Forbes Field, Building 321
Topeka, KS 66620
913-862-9360, Ext. 290

Table 2.1 State and Territorial Hazardous
Waste Management Agencies (Continued)

Kentucky
Division of Waste
Management
Department of Environmental
Protection
Cabinet for Natural Resources
and Environmental Protection
Fort Boone Plaza, Building #2
18 Riley Road
Frankfort, KY 40601

Louisiana
Hazardous Waste Division
Office of Solid
and Hazardous Waste
Louisiana Department
of Environmental Quality
P.O. Box 44307
625 N. 4th Street
Baton Rouge, LA 70804
504-342-9079

Maine
Bureau of Oil and Hazardous
Materials Control
Department of Environmental
Protection
State House Station #17
Augusta, ME 04333
207-289-2651

Maryland
Hazardous and Solid Waste
Management Administration
Maryland Department of the
Environment
201 W. Preston Street
Room 212
Baltimore, MD 21201
301-255-5647

Massachusetts
Division of Solid and Hazard-
ous Waste
Massachusetts Department of
Environmental Protection
One Winter Street, 5th Floor
Boston, MA 02108
617-292-5589

Michigan
Waste Management Division
Environmental Protection
Bureau
Department of Natural Re-
sources
Box 30038
Lansing, MI 48909
517-373-2730

Minnesota
Solid and Hazardous Waste
Division
Minnesota Pollution Control
Agency
520 Lafayette Road, North
St. Paul, MN 55155
612-296-7282

Mississippi
Division of Solid and
Hazardous Waste Management
Bureau of Pollution Control
Department of Natural Re-
sources
P.O. Box 10385
Jackson, MS 39209
601-961-5062

Missouri
Waste Management Program
Department of Natural
Resources
Jefferson Building
205 Jefferson St. (13/14 Floor)
P.O. Box 176
Jefferson City, MO 65102
314-751-3176

Montana
Solid and Hazardous Waste
Bureau
Department of Health and En-
vironmental Sciences
Cogswell Building
Room B-201
Helena, MT 59620
406-444-2821

Nebraska
Hazardous Waste Management
Section
Department of Environmental
Control
State House Station
P.O. Box 94877
Lincoln, NE 68509
402-471-2186

Nevada
Waste Management Program
Division of Environmental
Protection
Department of Conservation
and Natural Resources
Capitol Complex
201 South Fall Street
Carson City, NV 89710
702-687-4670

New Hampshire
Division of Public Health
Services
Office of Waste Management
Department of Health and Wel-
fare
Health and Welfare Building
6 Hazen Drive
Concord, NH 03301
603-271-4662

New Jersey
Division of Waste
Management
Department of Environmental
Protection
401 East State Street (CN 028)
Trenton, NJ 08625

New Mexico
Hazardous Waste Section
Groundwater and Hazardous
Waste Bureau
New Mexico Health and
Environment Department
P.O. Box 968
Sante Fe, NM 87504-0968
505-827-2924

Table 2.1 State and Territorial Hazardous
Waste Management Agencies (Continued)

New York
Division of Solid
and Hazardous Waste
Department of Environmental
Conservation
50 Wolfe Road, Room 209
Albany, NY 12233
518-457-6603

North Carolina
Solid and Hazardous Waste
Management Branch
Division of Health Services
Department of Human Re-
sources
P.O. Box 2091
Raleigh, NC 27602
919-733-2178

North Dakota
Division of Hazardous Waste
Management
Department of Health
1200 Missouri Avenue
Room 302
Box 5520
Bismark, ND 58502-5520
701-224-2366

Ohio
Division of Solid and Hazard-
ous Waste Management
Ohio Environmental Protection
Agency
1800 Watermark Drive
P.O. Box 1049
Columbus, OH 43266-0149
614-271-5338

Oklahoma
Waste Management Service
Oklahoma State Department
of Health
P.O. Box 53551
1000 Northeast 10th Street
Oklahoma City, OK 73152
405-271-5338

Oregon
Hazardous and Solid Waste
Division
Department of Environmental
Quality
811 Southwest 6th Avenue
Portland, OR 97204
503-229-5356

Pennsylvania
Bureau of Waste Management
Pennsylvania Department of
Environmental Resources
P.O. Box 2063
Fulton Building
Harrisburg, PA 17120
717-787-9870

Puerto Rico
Environmental Quality Board
Santurce, PR 00910-1488
809-725-0439

Rhode Island
Solid Waste Management
Program
Department of Environmental
Management
204 Cannon Building
75 Davis Street
Providence, RI 02908
401-277-2797

South Carolina
Bureau of Solid and Hazardous
Waste Management
Department of Health
and Environmental Control
2600 Bull Street
Columbia, SC 29201
803-758-5681

South Dakota
Office of Air Quality and Solid
Waste
Department of Water and Nat-
ural Resources
523 E. Capitol
Foss Building, Room 416
Pierre, SD 57501
605-773-3153

Tennessee
Division of Solid Waste
Management
Tennessee Department
of Public Health
701 Broadway
Customs House, 4th Floor
Nashville, TN 37219-5403
615-741-3424

Texas
Hazardous and Solid Waste
Division
Texas Water Commission
P.O. Box 13087
Capitol Station
Austin, TX 78711-3087
512-463-7760

Utah
Bureau of Solid and Hazardous
Waste Management
Department of Health
P.O. Box 16700
288 North 1460 West Street
Salt Lake City, UT 84116-0700
801-533-4145

Vermont
Waste Management Division
Agency of Environmental
Conservation
103 South Maine Street
Waterbury, VT 05676
802-244-8702

Virgin Islands
Department of Conservation
and Cultural Affairs
P.O. Box 4399, Charlotte
St. Thomas, VI 00801
809-774-6420

Virginia
Division of Technical Services
Virginia Department of Waste
Management
Monroe Building, 11th Floor
101 North 14th Street
Richmond, VA 23219
804-225-2667

Table 2.1   State and Territorial Hazardous
Waste Management Agencies (Continued)

Washington	Wisconsin	Wyoming
Solid and Hazardous Waste Management Division Department of Ecology Mail Stop PV-11 Olympia, WA 98504 206-459-6316	Bureau of Solid Waste Management Department of Natural Resources P.O. Box 7921 Madison, WI 53707 608-266-1327	Solid Waste Management Program State of Wyoming Department of Environmental Quality 122 West 25th Street Herschler Building Cheyenne, WY 82002 307-777-7752
West Virginia Waste Management Division Department of West Virginia Natural Resources 1260 Greenbriar Street Charleston, WV 25311 304-348-5935		

Reprinted from EPA, *Does Your Business Produce Hazardous Waste?* 1990, EPA/530-90-027.

The lead agency concept was developed so that when numerous agencies are involved, decisions could hopefully be made more quickly and efficiently. The generator would also deal primarily with one agency. This is a great concept, but rarely works as well as it could.

In terms of California, for example, there are five agencies that commonly get involved with HMW. For air toxics, it is usually the California Air Resource Board (ARB) or the Regional Air Quality Management Districts. For water toxics, it is usually the Regional Water Quality Control Boards. The California EPA regulates hazardous waste issues. The California EPA was started on July 17, 1991, and includes the State Water Resource Control Board (SWRCB), ARB, Department of Toxic Substances Control, Department of Pesticide Regulation, the Integrated Waste Management Board and the Office of Environmental Hazard Assessment. It is an umbrella of agencies.

Other Agencies of Importance

Many other agencies are involved in HMW to varying degrees. These agencies have toxics as only a small part of their overall purpose. This should be considered when dealing with these agencies. In certain situations, they may even act as the lead agency; however, they usually support the lead agency. For example, the local planning commission, city and county may get involved. A few examples of California agencies include the Agricultural Commission, Office of Emergency Services, Office of

Planning and Research, Fire Departments, Department of Water Resources, State Board of Equalization, Occupational Safety and Health Administration (California OSHA), California Highway Patrol (CHP) and State Fire Marshall. Other federal agencies of importance include the Department of Fish and Game, Department of Defense, Coast Guard, Department of Health and Human Services and federal OSHA. Local agencies of importance include water districts, air districts, fire, building, public works, planning, transit and health departments.

There is so much HMW legislation that it would be impossible to present it all in this book. A sampling of some of the major acts is shown in Table 2-2. As can be seen from this table, most states have some form of a hazardous waste management act. Most also have an underground storage tank type of law. Most of the legislation shown in Table 2-2 would have regulations that have been developed to implement the act. Sometimes it takes several years for the draft and final regulations to be printed, and then several more years for court cases to determine how the regulations should be interpreted. The rest of this chapter will concentrate on federal legislation and one state as an example, California.

Table 2.2   Examples of State Hazardous Material
and Waste Laws and Acts

Alabama Hazardous Wastes Management Act Alabama Underground Storage Tank Law	**Connecticut** Hazardous Waste Law
Arizona Hazardous Waste Disposal Law Underground Storage Tanks Act	**Delaware** Hazardous Waste Management Act Underground Storage Tank Act Hazardous Substance Cleanup Act
Arkansas Hazardous Waste Management Act Remedial Action Trust Fund Act Federally Listed Hazardous Sites Law Underground Storage Tanks Act	**Florida** Solid and Hazardous Waste Management Act Statewide Multipurpose Hazardous Waste Facility Siting Act
California Hazardous Substance Account Act Underground Storage of Hazardous Substances Act Hazardous Waste Control Act	**Georgia** Hazardous Waste Management Act Hazardous Waste Management Authority Act Underground Storage Tank Act Oil or Hazardous Material Spill or Release Act
Colorado Hazardous Waste Act Hazardous Waste Cleanup Act Underground Storage Tanks Law	**Hawaii** Underground Storage Tanks Act Hazardous Waste Management Act

Table 2.2 Examples of State Hazardous Material
and Waste Laws and Acts (Continued)

Idaho
Hazardous Waste Management Act
Hazardous Waste Facility Siting Act

Indiana
Hazardous Waste Act
Underground Storage Tanks Act
Hazardous and Solid Waste Landfills Act
Hazardous Waste Land Disposal Tax Act

Iowa
Hazardous Waste Facility Siting and Land Disposal Act
Underground Storage Tanks Act
Hazardous Waste Disposal Penalty Law
Infectious Waste Management Act

Kansas
Hazardous Waste Management Act
Superfund Law
Underground Storage Tank Act

Louisiana
Hazardous Waste Control Law

Maine
Hazardous Waste, Septage and Solid Waste Management Act
Underground Storage Facilities Law
Uncontrolled Hazardous Substance Sites Law
Hazardous Waste and Waste Oil Act

Maryland
Hazardous Waste Facility Siting Law
Hazardous Substances Spill Response Law

Massachusetts
Hazardous Waste Law
Oil and Hazardous Material Release Prevention and Response Act
Toxics Use Reduction Act

Michigan
Hazardous Waste Management Act
Underground Storage Tank Act

Minnesota
Underground Storage Tank Act
Environmental Liens for Cleanup Action Expenses Act
Toxic Pollution Prevention Act
Infectious Waste Control Act

Mississippi
Hazardous Waste Facility Siting Act

Missouri
Hazardous Waste Management Law
Underground Storage Tank Act

Montana
Hazardous Waste Act
Superfund Cooperation Act
Hazardous Substance Remedial Action Law
Underground Storage Tanks Act

Nebraska
Underground Storage Tank Act
Hazardous Waste Disposal Law

New Hampshire
Hazardous Waste Law
Underground Storage Facilities Act

New Jersey
Hazardous Waste Facilities Siting Act
Underground Storage Tanks Act

New Mexico
Hazardous Waste Act
Hazardous Waste Feasibility Study Act

New York
Solid and Hazardous Waste Management Law
Hazardous Substances Bond Storage Act
Hazardous Waste Bond Act

North Carolina
Solid and Hazardous Waste Management Act
Leaking Underground Storage Tanks Act

North Dakota
Hazardous Waste Management Act
Petroleum Release Compensation Act

Table 2.2 Examples of State Hazardous Material
and Waste Laws and Acts (Continued)

Ohio	**Tennessee**
Solid and Hazardous Waste Disposal Law	Hazardous Waste Management Act
Underground Storage Tank Law	Petroleum Underground Storage Tank Act
Oklahoma	**Texas**
Controlled Industrial Waste Disposal Act	Underground Storage Tanks Act
Underground Storage Tank Act	Toxic Chemical Release Reporting Act
Oregon	**Utah**
Hazardous Waste Management Act	Hazardous Waste Facility Siting Act
Underground Storage Tanks Act	Solid and Hazardous Waste Act
Spill Response and Cleanup of Hazardous	Hazardous Waste Facilities Management Act
Materials Act	Hazardous Substances Mitigation Act
Notice of Environmental Hazardous Act	Superfund Law
Hazardous Waste Remedial Action Act	
	Vermont
Pennsylvania	Hazardous Waste Generation Tax Law
Hazardous Sites Cleanup Act	Underground Liquid Storage Tank Law
Storage Tank and Spill Prevention Act	
	Virginia
Rhode Island	Hazardous Waste Management Act
Hazardous Waste Management Act	Underground Storage Tanks Act
Hazardous Waste Management Facilities Act	Hazardous Materials Emergency Response Act
Underground Storage Tanks Act	
Hazardous Waste Cleanup Act	**Washington**
Hazardous Waste Reduction Act	Hazardous Waste Management Act
Hazardous Substances Act	Model Toxics Control Act
	Hazardous Substance Tax Act
South Carolina	
Hazardous Waste Management Act	**West Virginia**
Underground Storage Tanks Act	Hazardous Waste Management Act
Infectious Waste Management Act	Hazardous Waste Emergency Response
	Fund Act
South Dakota	Underground Storage Tank Act
Hazardous Waste Management Act	
Underground Storage Tank Act	**Wisconsin**
Superfund Law	Hazardous Waste Management Act
	Underground Storage Tank Act

Note: This is not a complete listing.

Under each of the following headings are listed a few HMW laws
or regulations. These are, in the author's estimation, the primary laws and
regulations of importance. Many more could be listed that have some
impact. The headings follow roughly the life cycle of an HMW as presented in Figure 1-2.

Manufacture of Hazardous Materials

Toxic Substances Control Act (TSCA)

TSCA concentrates on the regulation of raw materials rather than finished products or process waste. The primary goal of TSCA is to regulate the introduction and use of new hazardous chemicals in two broad categories, chemical substances and mixtures. TSCA is administered primarily by the U.S. EPA. It imposes specific and detailed requirements on the use, storage and disposal of many specific chemicals such as polychlorinated biphenyls (PCBs), fully halogenated chlorofluoroalkanes, asbestos, hexavalent chromium and others. It was enacted in 1976 and provides requirements and authorities for identification and control of toxic chemical hazards to humans and the environment. TSCA establishes many programs to gather toxicity information, risk assessments and control actions. Based on the findings, EPA can require labels, bans and special handling. Pesticides, tobacco, firearms and ammunition, food and additives, drugs and cosmetics are exempt from TSCA. TSCA can be found at 15 United States Code 2601 to 2692 with regulations in 40 CFR.

PCBs were identified in TSCA as a special case. This included immediate regulation and phased withdrawal from the marketplace. Total enclosure, labeling and safe disposal were also specified. EPA developed rules on inspections, storage for use and disposal, authorized use of transformers and record-keeping concerning PCBs.

TSCA has specific testing requirements. EPA can require a manufacturer of a chemical to test for health and environmental effects if sufficient data does not exist or there is unreasonable risk. A committee of various agencies designates chemicals that need to be tested [Wentz 1989].

Introduction of new chemical substances requires specific actions. A 90-day advance filing of a premanufacture notification with the EPA is required. This shows intent to manufacture or import a new chemical (new meaning not on the list of chemicals published by the agency). New significant use is treated as a new chemical.

TSCA regulates more on a chemical-by-chemical basis, while RCRA regulates entire waste streams (containing multiple chemicals). TSCA controls a chemical's manufacture, importation, exportation, processing, distribution, use and disposal. Controls may include bans, labeling, record-keeping, testing, notices, recalls and limitations on use. Civil penalties for noncompliance can be as high as $20,000/day of violation. Criminal penalties for "knowingly and willfully" violating the statute can include up to $25,000/day of violation and up to one year of imprisonment or both [Allegri 1986].

Federal Insecticide, Fungicide, and Rodenticide Act (FIFRA)

FIFRA regulates pesticide registration, worker safety, classification, manufacture, storage, labeling, use and disposal. Testing and specific application (general and restricted use) registrations are required. The law is administered by the EPA. Dealers have to be licensed and applicators certified. The main orientation of FIFRA is toward groundwater protection and safety. FIFRA can be found in United States Code 136–136y with regulations in 40 CFR 152 to 186.

Shipment of Hazardous Materials

Hazardous Material Transportation Act (HMTA)

The HMTA regulations can be found in 49 CFR parts 100–177. It is administered by the DOT and controls the shipping of HMW. Chemical names, container certification, placards, labels and paperwork are key components or requirements. For example, if a hazardous material is being transported, a bill of lading is required; if a hazardous waste is being transported, a hazardous waste manifest is required.

Selecting the proper shipping name is essential in HMW transportation. The hazardous material tables in 49 CFR parts 100–177 must be followed to convert the common name to the proper shipping name.

Labeling and packaging requirements must be strictly followed. The labeling and packaging requirements specifically depend on the quantity and hazard classification of the chemical.

On January 1, 1991, the new DOT/United Nations Performance Oriented Packaging standards (POPs) were phased in. This relates to anyone shipping hazardous materials. Increased requirements are specified in terms of packaging, shipping papers, labeling and placarding. The whole purpose of the POPs was to bring the DOT's regulations into alignment with international requirements. The standards are to be fully implemented by October 1, 1996. As part of POPs, the DOT proposed a rule under DOT docket HM 181, which has been in effect since October 1, 1993. HM 181 will simplify and reduce transportation regulations and provide greater flexibility in packaging requirements due to new technology.

Oil Pollution Act of 1990

Transportation of petroleum products is covered by the Oil Pollution Act of 1990, which can be found in 33 United States Code 2701-2761. Also, SB 2040 (Keene) was signed by the California Governor on September 22, 1990 (California Code 8574.1-8670.72). This bill creates a state oil spill contingency plan that applies to the transportation, storage and processing of this hazardous material. The bill also establishes the State Interagency Oil Spill Committee and Administrator for oil spill response. It requires that the Office of Emergency Services be notified concerning oil spills.

Hazardous Substances Highway Spill Containment and Abatement Act

Senate Bill 921 (Vuich) was approved by the California Governor on August 2, 1989, and also applies to the transportation of hazardous materials (California Vehicle Code 2450-2454). This bill requires that a toxic disaster contingency plan be prepared and that the CHP is responsible for scene management when hazardous materials are spilled during transportation.

California Highway Patrol (CHP) Information Bulletin

The CHP Information Bulletin can be found in the California regulation Title 13 (Section 1150). It covers various HMW shipment requirements such as driver training, licensing, shipping names, safety, maintenance, paperwork, labeling and placarding. The CHP also helps to enforce route and time restrictions. Annual inspections and certification of vehicles and containers is done by the CHP.

Atomic Energy Act of 1954

The Atomic Energy Act established the Atomic Energy Commission (AEC), which allowed only the federal government to regulate nuclear waste. The Energy Reorganization Act of 1974 abolished the AEC and divided its function between the Nuclear Regulatory Commission (NRC) and the Energy Research and Development Administration. Regulations can be found in Title 10 CFR, Energy. For example, the NRC was involved in the shipment of Three Mile Island waste to various research and disposal sites in several states.

Various Other State and Local Regulations

Many state and local agencies have notification, routing, permit and licensing requirements that apply to HMW haulers. Some local agencies also require registrations. Once a shipment of extremely hazardous material leaves the interstate highway, the local sheriff or police may also be involved.

Use and Storage of Hazardous Materials

Use and storage of hazardous material is covered by TSCA, which was previously discussed, SARA, RCRA Subtitle I (underground storage of hazardous materials), OSHA, UFC and the following California legislation: AB 2185, Prop. 65, Sher Bill, AB 3777 and the Above Ground Storage Act.

Superfund Amendments and Reauthorization Act (SARA) Title III

SARA Title III is a federal law (1986) that establishes state/local emergency response notification procedures and stored chemical notifications (community right to know). It is an amendment to and interface with the Federal Superfund Act and establishes cleanup standards.

The essential SARA Title III requirements for generators include chemical inventory, emergency planning, toxic chemical discharge and MSDS considerations. Inventory data must be submitted to the local agency by facilities that handle more than the threshold quantities (ranging from 1 lb. to 10,000 lbs.) of any of the 366 extremely hazardous substances. Unauthorized discharges of any extremely hazardous substance over the reportable quantity (RQ) must be reported. A list (indicating maximum daily usage and locations by hazard class) of all on-site chemicals requiring an MSDS must be provided to the agency (Tier I). Upon request, the actual MSDS must be provided (Tier II). In addition, the agency must receive an annual chemical inventory for all chemicals requiring an MSDS if more than 10,000 lbs. of OSHA-designated hazardous substances are stored on site (500 lbs. for extremely hazardous substances) [Elliott 1990]. The data submitted is available to the public on a computer database monitored by the EPA.

AB 2185/Community Right to Know/The Waters Bill

AB 2185 allows counties or communities in California to know the types of toxic chemicals and hazardous materials in their areas (California Health and Safety Code 550-25520). Companies must inventory chemicals and hazardous materials that are used or stored over certain volumes. The company must also develop notification and training procedures. Such procedures are known as Hazardous Material Business Plans (HMBP). Depending on the agency, the cut off volume for reporting varies; however, it is usually at or above 55 gallons, 500 lbs. or 200 cubic feet for any hazardous material or hazardous waste. This information must be submitted to the appropriate agency and is similar to SARA Title III, except at the state level.

Safe Drinking Water and Toxic Enforcement Act of 1986/Prop. 65

This referendum was passed by the voters of California to provide protection against chemicals causing cancer, birth defects or other reproductive harm (California Health and Safety Code 25249.5-25249.13). The Governor must publish lists of chemicals known to the state to cause these effects. The Office of Environmental Health and Hazard Assessments is the administering agency. Proof of compliance rests with the generator of the chemical or user. The act includes a bounty hunter provision.

Any organization (other than agencies) with 10 or more employees that have listed chemical(s) on site, which pose significant risk to employees or the public, must:

- Warn within 12 months after the chemical appears on the list and quarterly thereafter. Warnings may be done by labels, mailing notices, media, signs and employee hazard communication programs.

- Stop discharge of the chemical within 20 months after the chemical appears on the list if the chemical may enter a drinking water source.

Sher Bill

In California, the 58 counties or 42 cities (Fire Departments) regulate underground hazardous substance storage to prevent groundwater contamination. This applies to both existing and new tanks. Monitoring systems, permits and reporting of unauthorized releases are required (California

Health and Safety Code 25280-25299.7). In addition to the Sher Bill, the Cortese Bill, the HSWA of 1984 and many local ordinances (such as the Santa Clara County Hazardous Material Ordinance) also regulate underground storage tanks.

Acutely Hazardous Materials Risk Management

California Health and Safety Code 25531-25541 allows counties, cities or fire departments to require a Risk Management and Prevention Program (RMPP) if the organization is storing or handling acutely hazardous materials above listed thresholds. It requires plans if there is a significant likelihood that the material may pose an acutely hazardous materials accident risk. The RMPP includes hazardous material storage requirements, emergency response planning and the public right to know. An acutely hazardous material is approximately equal to an extremely hazardous material.

Aboveground Petroleum Storage Act

On October 2, 1989, SB 1050 (Torres) was signed into law (California Health and Safety Code 25270-25270.13). This requires California's Regional Water Quality Control Board (RWQCB) to inspect aboveground crude oil storage tanks. Each owner must prepare a spill prevention control and countermeasure plan (SPCCP), install monitoring systems, warn agencies of a release and submit hazardous substance storage statements. This is in addition to the EPA requirements for an SPCCP for aboveground storage of over 42,000 gallons or more of petroleum product.

Occupational Safety and Health Act (OSHA)

The federal OSHA has generated limitations for toxic and hazardous substance safety in the workplace. These regulations are found in Title 29 and include minimum standards for the prevention of harmful exposure of employees to dusts, fumes, mists, vapors and gases. Most of these standards are expressed as Permissible Exposure Levels (PELs) and are calculated as Time Weighted Averages (TWAs). New PELs are added from time to time. OSHA generally allows six to nine months to meet new standards. Threshold limit values (TLVs) identify the airborne concentration of a material to which a person can be exposed day after day without adverse effect.

Other important OSHA considerations include confined space restrictions, listing of hazardous substances and processes, regulated carcinogens, fumigation procedures and restrictions, and labeling of injurious substances. OSHA also requires that employers must maintain up-to-date health records on exposed employees and communicate hazards via a written hazard communication plan.

On May 5, 1992, the federal OSHA's Bloodborne Pathogen Standard became effective, which requires organizations to identify biological hazards. An exposure control plan, training and hepatitis B vaccinations must be provided to affected employees.

Certain states have been authorized to enforce OSHA regulations. California is one such state. In these states, the state regulations must be at least as stringent as the federal. Therefore, by complying with the state regulations you will be complying with the federal regulations. This process is also referred to as "delegation of enforcement authority."

Effective July 1, 1991, a new Cal/OSHA Injury and Illness Prevention Program went into effect as a result of SB 198. This required all employers (all sizes and industries) to have a written program.

Uniform Fire Code (UFC)

The fire departments specify construction standards for fixtures and buildings in order to minimize fire and explosion damage. The safe use, storage, and handling of hazardous materials are emphasized. The Hazardous Material Management Plan (HMMP) is also prepared in reference to this code. The HMMP provides data to the fire department concerning the types and locations of chemicals before a fire breaks out.

The UFC identifies safeguards for specific industries and processes where hazards to employees, fire fighters and the public may exist. Part IV—Special Occupancy Uses, Part V—Special Process, Part VI—Special Equipment and Part VII—Special Subjects are all broken down into articles that cover specific hazardous material situations.

Generation of Hazardous Waste

Hazardous waste generation is controlled by RCRA, HWCL, CWA, the Clean Air Act, the Water Quality Control Act, and, to a lesser extent, by Prop. 65, which was discussed earlier.

Resource Conservation and Recovery Act (RCRA)

RCRA, with its amendments, is possibly the single most important hazardous waste legislation. The primary goal of RCRA is to protect water, land and air from contamination by solid waste. A secondary goal is to control hazardous waste and set standards for the identification, listing and handling of hazardous waste. RCRA covers hazardous waste from the generator through final disposal.

RCRA is similar to Europe's Duty of Care Act. RCRA requires a manifest, where in Europe a "Consignment Note for the Carriage and Disposal of Special Waste" is required.

RCRA is administered by the EPA and can be found in 40 CFR parts 261, 264 and 268. RCRA amends the Solid Waste Management Act of 1965 and the Resource Recovery Act of 1970. RCRA regulates current and future hazardous waste disposal activities, while CERCLA regulates past disposal activities. RCRA deals primarily with hazardous waste *streams* from their point of origin to the point of disposal.

RCRA has been expanded three times since its passage. The first amendment occurred in 1980. In 1984, RCRA was upgraded by the Hazardous and Solid Waste Amendments. This added hammer provisions in reference to land bans and treatment standards, Class I specifications and other upgrades. HSWA broadened the scope of RCRA, especially in terms of ground water protection. In 1989, RCRA was also expanded and again, in 1992, with the Federal Facility Compliance Act.

Presently, RCRA does not cover most petroleum and mining-produced wastes, such as refuse from the coal mining industry. When and if these wastes are designated hazardous under RCRA, the amount of hazardous waste in the U.S. requiring special handling will probably increase by more than 300%.

Authorization has been given for most states to administer their own programs. California has been authorized to administer its own plan under the Hazardous Waste Control Law (HWCL) and the Porter-Cologne Water Quality Control Act.

Two subtitles in RCRA are of paramount importance, C and I. Subtitle C provides the hazardous waste management program framework and regulates hazardous waste from cradle to grave. Key provisions include listing of hazardous waste, delisting, EPA ID numbers, regulation of hazardous waste underground storage tanks, and manifests. Subtitle I regulates underground storage tanks for products.

RCRA is closely related to Superfund, which is discussed in Chapter 8. There is a certain amount of overlap, as illustrated in Figure 2-2. Both Superfund- and RCRA-generated wastes are treated at RCRA-approved facilities. On the far side of Figure 2-2 is a factory. This refers to the RCRA-permitted generator. Their waste, and that from Superfund sites and

Figure 2.2 Relationship of RCRA to Superfund. (Reprinted from EPA, *Comparing RCRA and Superfund*, Washington.)

spills, are hauled by an RCRA-permitted transporter to an RCRA-permitted treatment, storage, or disposal facility (TSDF).

The Solid Waste Enforcement Act of 1990 is new federal legislation that enlarges the authority of RCRA. Approximately 25 RCRA constituent chemicals are added to the priority pollutant list. Other pollutant levels are lowered, resulting in more wastes being handled by TSDF. The act also requires that the Toxicity Characteristic Leaching Procedure (TCLP) is utilized.

The new TCLP was promulgated by EPA on March 29, 1990. This approximately tripled the amount of waste considered hazardous under RCRA. It also impacted up to 15,000 more generators. The TCLP replaced the Extraction Procedure Toxicity Test (EP Tox) on September 25, 1990, for large-quantity generators and on March 29, 1991, for small-quantity generators. The TCLP is a more aggressive leaching procedure than the EP Tox Test.

A copy of RCRA and RCRA regulations can be obtained by calling the Government Printing Office. If the entire 40 CFR is not available, at least parts 260 through 271 should be ordered.

Hazardous Waste Control Law (HWCL)

The HWCL is very similar to RCRA and was written to implement RCRA. It can be found in California Health and Safety Code 25100-25249 with regulations in Title 22 and 23 of the California Code of Regulations. EPA and California DTSC have entered into an agreement that gives the DTSC the authority to implement the Title 22 and 23 regulations. Examples of important provisions include hazardous waste facility permits, hazardous waste hauler registration, requirements for generators of hazardous wastes, fees, hazardous waste and hazardous material lists, criteria for identification of hazardous and extremely hazardous wastes, recyclable hazardous wastes, infectious wastes, land disposal restrictions, general facility standards, manifests, treatment, annual registration, equipment certification and class I and II site standards.

Clean Water Act (CWA) and Federal Water Pollution Control Act

The CWA authorizes the EPA and RWQCB to control the discharge of harmful quantities of oil or toxic pollutants into U.S. waters. It requires a National Pollutant Discharge Elimination System (NPDES) permit for discharges into U.S. waters and establishes discharge standards. The CWA regulations can be found in 40 CFR parts 109 to 130.

The major goals of the CWA include (1) Restore and maintain the integrity of U.S. waters; (2) Restore to fishable and swimmable conditions; (3) Establish 126 priority pollutants; (4) Establish best available control technology (BACT) and best practical control technology (BPCT). BPCT is really a technology-based standard. It primarily applies to point sources (other than POTWs).

The Federal Water Pollution Control Act was passed in 1972 and amended in 1977 and 1981, and is administered by the EPA and local publicly owned treatment works (POTW) and requires best available technology (BAT) for toxic pollutants. It specifies pretreatment standards and effluent standards for discharge to POTWs and to any surface water. Under this act, only very limited amounts of hazardous waste are allowed to be discharged and in strict compliance with permit specifications (NPDES and local POTW permits). An NPDES permit is also required for stormwater runoff if it is impacted by HMW.

Clean Air Act

The Clean Air Act establishes the national emission standards for hazardous air pollutants. The seven major air quality hazards specified under the Clean Air Act (CAA) are total suspended particulates, sulfur dioxide, carbon monoxide, nitrogen dioxide, ozone, lead and hydrocarbons. An additional five hazardous air pollutants have been added: asbestos, benzene, beryllium, mercury and vinyl chloride. A Prevention of Significant Deterioration (PSD) permit is required for some discharges to the air to maintain the air quality standards for attainment areas. Many aspects are implemented by the state air agency, such as California's Air Resource Board of local air quality management districts. The Clean Air Act can be found in 42 United States Code 7401-7642 with regulations in 40 CFR parts 50-87. The Clean Air Act has a set of technology-based standards. These include best available control technology (BACT) for new sources of air pollutants.

The Clean Air Act (CAA) Amendments of 1990 existed in the Senate and House bill stage for about three years, and each year were modified and fine-tuned. The bill was passed into law in 1990. One of the requirements of the law is that the EPA must now develop regulations to implement the provisions of the act.

Of particular interest in the amendments are the hazardous air pollutants (HAPs). In 1970, the original CAA gave the EPA authority to list and regulate HAPs. The EPA only listed eight substances. The 1990 amendments statutorily identified 189 such substances. This forces the EPA to develop regulations for considerably more HAPs than they have ever dealt with.

Short-Term Storage of Hazardous Waste

Hazardous waste storage is regulated under some of the same regulations discussed previously, which include RCRA, HWCL, Community Right to Know and OSHA.

Shipment of Hazardous Waste

Shipment of hazardous waste is controlled by regulations discussed previously, which include HMTA, RCRA, HWCL, the CHP Information Bulletin and the Nuclear Regulatory Commission.

Long-Term Treatment, Storage and Disposal

Long-term treatment, storage and disposal is covered by RCRA, HWCL and TSCA, which were presented earlier.

In addition, various California legislations cover these activities. The Tanner Bills (AB 1807/2948) are administered by the counties and cover disposal sites. The Katz Bills (AB 3566/3121) are administered by the RWQCB and restrict the use of pits. The Calderon and Eastin Bills are administered by the RWQCB and the AQMD and cover air, leachate testing and Superfund matters. The Roberti Bill is administered by California EPA and applies to land disposal phase-out. The Tanner Bill is administered by the county and applies to household hazardous waste disposal. SB 1469 was passed in 1992 and increases generator/manifest fees.

The Wright-Lempert Hazardous Waste Treatment Permit Reform Act was passed in California in 1992 (AB 1772). This allows for tiered hazardous waste treatment permitting in California. The different possibilities include full RCRA permit, standardized permit for off-site commercial facilities handling California hazardous waste, permit-by-rule for on-site treatment, and conditionally authorized on-site treatment and conditionally exempt for on-site treatment. The bill extensively revised the HWCL.

In summary, the regulations presented in this chapter have concentrated on "normal" HMW activities. The special or problem situations are covered by a different set of regulations, which are discussed in Chapter 8, such as the Comprehensive Environmental Response, Compensation and Liability Act.

Documents Required by Agencies

One way for agencies to determine compliance and to collect fees is to require permit applications, various plans and documents of organizations. In addition, these force the organization to do considerably more advanced planning than they might have ordinarily done. Unfortunately, many of the applications and other pieces of paper are repetitive and sometimes conflicting.

EPA Identification Number

Although not really a document, the EPA requires an EPA ID number of all hazardous waste generators. This number is referenced on numerous documents, including the manifest.

In general, the EPA ID number is required prior to transport and disposal or recycling of hazardous wastes. Small-time generators, such as households, do not need an EPA ID number; however, most commercial and industrial organizations would. There are some small-volume exemptions, which vary depending on the waste.

An EPA ID number (and manifesting) is usually required even if the hazardous waste is to be recycled. If treatment of the waste is not required prior to reuse or the waste is to be burned for energy recovery, then the number may not be required. In general, organizations (other than households) that generate a hazardous waste for disposal or that have used solvents, antifreeze or ink recycled would need an EPA ID number.

Environmental Reports

Depending on whether there are significant impacts involved, an Environmental Assessment (EA) and possibly an Environmental Impact Statement (EIS) for the federal agencies may be required in support of a permit application. An initial study (IS) and possibly an environmental impact report (EIR) may be required by California agencies according to the California Environmental Quality Act. In accordance with the National Environmental Policy Act of 1969 (NEPA), the documents are usually prepared under the direction of the lead agency with input from the applicant, consultants, other agencies and the public. The EA or IS is usually pre-pared first, and if no significant impacts are found, a negative declaration or finding of no significant impact is issued. If this does not occur, then an EIR and/or EIS is prepared. This process can take several years.

Permits and Variances

Various federal, state, regional and local agencies require permits in relationship to HMW activities. The California DTSC, for example, has a full range of permits and variances depending on the need and situation. These vary dramatically in terms of preparation and processing time for both the applicant and DTSC. They include (generally from least involved to most involved) exemptions, variances, permits by rule, standardized per-mits and full permits.

Authority to construct and permit to operate Permits are usually required for processes that create air pollution. They are adminis-tered by the state air agency, such as California's Bay Area Air Quality Management District, and can involve both hazardous and nonhazardous air pollutants. The authority to construct permit is for approval to start con-structing an air-quality-related project, and the permit to operate is to allow operation of the air-emittion equipment.

National pollutant discharge elimination system (NPDES)
An NPDES permit is required if there are discharges to U.S. waters. The permit is usually issued by the state agency, such as California's SWRCB or the RWQCB. Traditionally, an NPDES permit was required for a point source or end-of-pipe-type industrial discharge. NPDES permits are now also required for nonpoint storm water run-off from industrial and other areas believed to cause water pollution.

Hazardous waste facility permit A hazardous waste facility permit is also called a treatment, storage, or disposal facility (TSDF) per-mit. It is required if on-site hazardous waste activities include storage over 90 days, treatment, recycling, incineration, placement in surface ponds or disposal on land.

TSDFs that were in operation on November 19, 1980, and that met certain conditions were given an "Interim Status." To obtain this status, an operator would have to notify the EPA about its operation prior to that date. The status continues until a final permit is issued or denied.

•HAZARDOUS WASTE FACILITY SUBMITS PERMIT APPLICATION*

•U.S. EPA AND STATE EVALUATE PERMIT APPLICATION*

•AGENCIES RECOMMEND APPROVAL OR DENIAL OF PERMIT*

•FACT SHEET AVAILABLE TO COMMUNITY**

•DRAFT PERMIT AND PERMIT APPLICATION AVAILABLE FOR PUBLIC REVIEW**

•DRAFT PERMIT OR DENIAL NOTICE IS PREPARED AND ISSUED*

•45-DAY PUBLIC COMMENT PERIOD**

•PUBLIC HEARING TO RECEIVE COMMENTS AND ADDRESS CONCERNS**

•PERMIT REVISED IF APPROPRIATE*

•PUBLIC COMMENT PERIOD ENDS*

•PREPARE RESPONSE TO PUBLIC COMMENTS*

•RESPONSE SUMMARY AVAILABLE**

•U.S. EPA AND STATE REACH DECISION*

•NOTICE OF DECISION*

•CONTINUED MONITORING AND INSPECTION OF FACILITY*

*PERMIT PROCESS
**PUBLIC INVOLVEMENT

Figure 2.3 Getting an RCRA Permit. (Source: EPA Region V and EPA, *The RCRA Permitting Process Fact Sheet.*)

The TSDF permit is usually administered by the state agency, such as California's DTSC. However, EPA has authority over all RCRA permits.The full-blown permit application is made up of Part A, which is a brief introduction, and Part B, which is a detailed operation plan. EPA excludes some facilities from needing a TSDF permit, such as certain hazardous waste reuse or recycle facilities, totally enclosed treatment facilities and a few other operations.

Obtaining this type of permit is normally a very long and involved process, as Figure 2-3 shows. There are even some activities not shown in Figure 2-3 that add additional time. For example, some up-front design work and agency interface must occur before the first item shown occurs, which is submittal of the permit application. One of the most important parts in the entire figure in terms of timing is "Draft Permit or Denial

Notice Is Prepared and Issued." This is the real critical point in the entire process. If the applicant established good communication with the agency, they would unofficially know the outcome long before this point in time.

Operations that hold a TSDF permit must comply with more requirements than a normal generator. Some of these extra considerations are shown in Table 2-3 and are compared to common generator requirements. For example, a TSD facility must prepare a closure plan, while a generator may not be required to do so unless a plan is required by a local regulatory agency.

Table 2.3 Generator Requirements and Treatment, Storage and Disposal Facility (TSDF) Requirements

	Generator	TSDF
Waste Determination	Can assume hazardous	Lab test required
Inspection Records	Not required	Required
Operating Records	Not required	Required
Report to State Agency	Biannual	Annual
Closure Plan	Usually not required by state or federal. May be local agency requirement.	Required
Storage	Usually less than 90 days	Greater than 90 days

A full-blown TSDF permit can sometimes be avoided altogether by altering the process. For example, if a hazardous material is "pulled out" of the process stream before it reaches the end and is reused, a TSDF permit may not be required. It is also possible to get a variance from the agency in certain cases to treat small volumes of hazardous waste in drums. A permit by rule (PBR) for a transportable treatment unit (TTU) is another way to avoid a TSDF permit. In this situation, if plans for a TTU are turned in for certain sites and certain wastes (such as heavy metals), the agency is often required to give a decision or permit within 30 days.

Conditional land use permit   A conditional land use permit is required if there is land use impact or commercial/industrial development. It is administered by the local agencies, such as counties, and applies to numerous operations, including those that involve both hazardous and non-hazardous waste.

Extremely hazardous waste disposal permit   An extremely hazardous waste disposal permit is required of a generator by some states when wastes contain substances such as arsenic, beryllium, lead and mercury in certain concentrations. These permits are issued for one-time use or for up to one year, usually by the state hazardous waste agency.

Fire permits   Fire permits are administered by local fire departments and are required for flammable material storage. These agencies inspect facilities for compliance with the Uniform Fire Code. In some cases, an official permit is not required or issued; however, inspections may still occur.

Underground tank permits   These permits are required of owners of underground tanks and other underground structures and are issued by the fire departments and county health departments. Monitoring systems are an important aspect of the permit application.

County public health license   A County Public Health License is required of some generators of hazardous waste in certain counties. There is wide variation between counties concerning quantities and types of waste that trigger the need for the license. The county will normally inspect at least once a year.

Industrial wastewater discharge permits   These are required of operations that discharge wastewater from a process into a publicly owned treatment works (POTW). It is administered by the local POTW and also regulated by federal categorical pretreatment standards. More and more of these "source control" type permits are being required.

Variances to ship certain hazardous wastes   The state hazardous waste regulator will, in certain cases, issue low-level and/or low-volume variances that allow organizations to transport hazardous wastes without manifests and without using licensed haulers.

Plans

Medical waste management plans   California's AB 1641 and AB 109 require medical facilities to file a medical waste management plan. This includes both small generators (less than 200 lbs./month) who treat on-site and large-quantity generators. Medical waste requirements are administered by state agencies or local county health departments. These agencies regulate on-site treatment (including autoclaves and microwaves), storage and disposal of medical waste.

Pollution prevention plan SB 1726 was passed in California in 1992 and requires that a Pollution Prevention Plan be submitted every four years with biennial reports. This applies to generators of over 5000 kg of hazardous waste. It lowered the reporting threshold of 10,000 kg specified in SB 14.

Hazardous material management plan (HMMP)

The HMMP pertains primarily to underground storage tanks and presents information such as types of tanks, leak detection and material stored. In the future, fire departments may refuse to enter a property to put out a fire unless there is an HMMP posted. In some cases, the local fire department may combine the fire permit with the HMMP.

A typical HMMP may contain the following sections: introduction, facility description, hazardous substance storage statement, separation of materials, monitoring program, record-keeping forms, emergency equipment, contingency plan, hazardous materials inventory statement, hazardous substance storage statement, monitoring program, data chart for tank system tightness test, hazardous material incident report form and vicinity and facility storage maps.

Hazardous material business plan (HMBP) An HMBP is a document containing hazardous material management information, site specific maps and chemical inventories. In California, AB 2185 and the Community Right to Know laws require the HMBP, which is administered by the fire department or county. These agencies issue a Certificate Of Disclosure upon approval of the HMBP. A typical HMBP can be composed of three sections, including inventory reporting information, emergency response plans and procedures, and employee training.

Both federal and state regulations require plans that address hazardous waste emergencies such as spills. Therefore, some operators prepare a separate "Hazardous Waste Contingency Plan." If, however, the operation already has the subjects covered in their HMBP or emergency response, this plan should meet regulatory requirements.

Risk management and prevention plan (RMPP)

An RMPP is required if a facility stores or uses a significant quantity of extremely or acutely hazardous materials. The emphasis in the RMPP is on emergency response planning. The RMPP is administered by counties, cities or fire departments. The statute can be found in Health and Safety Code section 25531-25541.

Hazard recognition/communication plan This plan pertains to training for field forces in safety matters, including recognition and safe handling of hazardous materials. The plan should be written by all organizations with over 10 employees to meet OSHA and other requirements.

A typical plan will contain a hazard recognition section and a hazard communication section. The recognition section presents types of hazards and how chemicals enter and affect the body. The potential health hazards at company locations and common hazards are also discussed. The communication section usually presents the written company program, how to identify a hazardous substance, labeling requirements, new product information and material safety data sheets.

Injury and illness prevention plans (IIPP) Another requirement of OSHA is the IIPP. Some of the components mentioned above in terms of hazard recognition are also part of this plan. This requirement went into effect on July 1, 1991. The intent of the IIPP is to identify all potential hazards in the work area and provide information on how to respond to any accidents. The plan also identifies all responsible individuals and the proper procedure to contact these parties.

Manifests and Exception Reports

The purpose of the uniform hazardous waste manifest (manifest) is to track and create a paper trail of the progress of a hazardous waste from point of generation to point of disposal. A manifest is required of any generator of a hazardous waste who transports the waste (or has it transported) or turns it over for treatment, storage or disposal.

A manifest (see Figure 2-4) must be completed perfectly, or injuries and environmental damage may occur. Also, the generator may be fined heavily. Errors in manifest preparation are one of the most common fines, especially in completing Section 11 (U.S. DOT Description). This description must be word-for-word as specified by DOT. If the description is put in wrong, there is a chance that the transporter may be injured or the environment contaminated on the way to the TSDF. The TSDF personnel might also be injured if they don't catch the error before they start handling the waste.

After the manifest is signed, the copies of the manifest go to various places (see Figure 2-5). Copy 6 is kept by the generator. Copy 5 goes to the state regulatory agency in many states. Copies 4, 3, 2 and 1 are given to the transporter, who signs and retains copy 4. The transporter gives copies 3, 2, and 1 to the disposal facility upon arrival. The disposal facility signs and

STATE OF ARKANSAS
Department of Pollution Control and Ecology
P. O. Box 9583 Little Rock, Arkansas 72219
Telephone 501-562-7444

EMERGENCY RESPONSE

1

Please print or type.   (Form designed for use on elite (12-pitch) typewriter.)

Instructions for completing this form
are on the back of Part 1.
Form Approved. OMB No. 2000-0404. Expires 7-31-86

UNIFORM HAZARDOUS WASTE MANIFEST	1. Generator's US EPA ID No.	Manifest Document No.	2. Page 1 of	Information in the shaded areas is not required by Federal law.

3. Generator's Name and Mailing Address		A.State Manifest Document Number **AR- 53109**
		B.State Generator's ID
4. Generator's Phone ( )		
5. Transporter 1 Company Name	6. US EPA ID Number	C.State Transporter's ID
		D.Transporter's Phone
7. Transporter 2 Company Name	8. US EPA ID Number	E.State Transporter's ID
		F.Transporter's Phone
9. Designated Facility Name and Site Address	10. US EPA ID Number	G.State Facility's ID
		H.Facility's Phone

	11.US DOT Description *(Including Proper Shipping Name, Hazard Class, and ID Number)*	12.Containers		13. Total Quantity	14. Unit Wt/Vol	I. Waste No.
	HM	No.	Type			
G E N E R A T O R	a.					
	b.					
	c.					
	d.					

J. Additional Descriptions for Materials Listed Above	K. Handling Codes for Wastes Listed Above

15. Special Handling Instructions and Additional Information

16. **GENERATOR'S CERTIFICATION:** I hereby declare that the contents of this consignment are fully and accurately described above by proper shipping name and are classified, packed, marked, and labeled, and are in all respects in proper condition for transport by highway according to applicable international and national governmental regulations , and Arkansas state regulations.

Printed/Typed Name	Signature	Date Month Day Year

T R A N S P O R T E R	17. Transporter 1 Acknowledgement of Receipt of Materials		Date
	Printed/Typed Name	Signature	Month Day Year
	18. Transporter 2 Acknowledgement or Receipt of Materials		Date
	Printed/Typed Name	Signature	Month Day Year

F A C I L I T Y	19. Discrepancy Indication Space		
	20. Facility Owner or Operator: Certification of receipt of hazardous materials covered by this manifest except as noted in Item 19.		Date
	Printed/Typed Name	Signature	Month Day Year

EPA Form 8700-22 (3-84)

NOTICE: THE ORIGINAL AND NOT LESS THAN TWO (2) COPIES MUST MOVE WITH THE HAZARDOUS WASTE SHIPMENT. ONCE DELIVERED, THE TREAT-MENT/STORAGE/DISPOSAL FACILITY MUST RETURN THIS ORIGINAL COPY TO THE GENERATOR.

Figure 2.4 State of Arkansas Uniform Hazardous Waste Manifest.

retains copy 3 and sends copy 2 to the regulatory agency and copy 1 to the generator. Both the agency and the generator will end up with two copies, one initiated at the beginning of the process and one at the end, allowing a match to occur. A missing copy is a sign that something may have gone wrong.

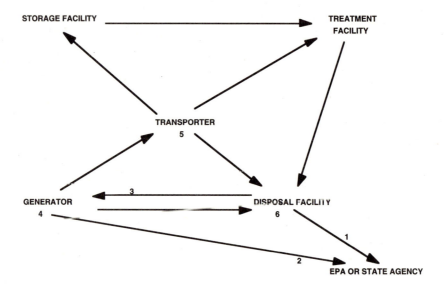

Figure 2.5 Tracking Waste "from Cradle to Grave." A one-page manifest must accompany every waste shipment. The result documents the waste's progress through treatment, storage, and disposal. A missing form alerts the generator to investigate, which may mean calling in the state agency or EPA. (Reprinted from EPA, 1993.)

Table 2-4 shows the documentation that must be sent along with a manifest. This additional paperwork relates to land ban or land disposal restriction (LDR) and, depending on the situation, includes notification and certification forms. Note in Table 2-4 that a waste can exceed LDR levels and still be shipped to a permitted TSDF with the proper notification form. If the waste exceeds LDR, it cannot be sent to a TSDF unless it has a variance and the disposal facility receives a notification of such.

The notification sent when a restricted waste exceeds treatment standards must include various items or the waste will be rejected. The EPA ID number must be sent along with the applicable treatment standard. In addition, the manifest number of the shipment and waste analysis data must also be sent.

Table 2.4 Paperwork that Must Accompany
Hazardous Waste Manifests

Waste Stream	Management Alternative	Paperwork	To Whom
LDR waste with treatment standards	Waste concentration exceeds LDR level. Generator sends to treatment or storage facility	Notification	Treatment or storage facility
	Waste concentration is below LDR level. Generator sends to hazardous waste landfill, treatment or storage facility	Notification and certification	Landfill, treatment or storage facility
LDR waste with treatment standards subject to variance	Management sends waste directly to hazardous waste landfill	Notification	Hazardous waste landfill

Reprinted from EPA, *RCRA Orientation Manual—1990 Edition* (Washington, 1990).

When the restricted waste meets the treatment standard without further treatment, a certification stating that the waste complies must be sent. Also, the same notification form mentioned above must be sent as well.

An exception report must be filed by the generator if the manifest is not received back from the disposal site within 45 days. Both the manifest and the exception report must be retained for at least three years.

The biggest difference between a manifest and a bill of lading, which will be discussed next, is the EPA ID number. A manifest requires an EPA ID number for the generator, transporter and disposal facility, where a bill of lading does not.

Bills of Lading

A bill of lading is required for the transportation of hazardous materials (where a manifest is for transporting hazardous waste). The bill of lading must accompany the load of hazardous material. Figure 2-6 is an example of a bill of lading.

The amount of detail required on a bill of lading is significantly less than that required on a manifest. In addition, there are fewer copies and less tracking required.

Biennial Reports

A report must be sent to the state regulatory agency concerning waste generation and minimization. The source of the data is from manifests. The

CONTAINS HAZARDOUS MATERIALS
FOR HELP IN CHEMICAL EMERGENCIES INVOLVING SPILL, LEAK,
FIRE OR EXPOSURE CALL TOLL-FREE 1-800-424-9300 DAY OR NIGHT

STRAIGHT BILL OF LADING
ORIGINAL - NOT NEGOTIABLE

Shipper's No. _____

Carrier's No. _____

(NAME OF CARRIER)   SCAC _____   Date _____

TO:
Consignee _____

FROM:
Shipper _____

Street _____   Street _____

Destination _____   Zip _____   Origin _____   Zip _____

Route: _____   Vehicle Number _____

No. Shipping Units	HM	Kind of Packages, Description of Articles (IF HAZARDOUS MATERIALS - PROPER SHIPPING NAME)	HAZARD CLASS	I.D. Number	WEIGHT (subject to correction)	RATE	LABELS REQUIRED (or exemption)

Remit C.O.D. to:
Address:
City: _____ State: _____ Zip: _____

COD Amt: $ _____

C.O.D. FEE:
Prepaid ☐
Collect ☐ $ _____

NOTE - Where the rate is dependent on value, shippers are required to state specifically in writing the agreed or declared value of the property. The agreed or declared value of the property is hereby specifically stated by the shipper to be not exceeding

$ _____ Per _____

Subject to Section 7 of the conditions, if this shipment is to be delivered to the consignee without recourse on the consignor, the consignor shall sign the following statement.
The carrier shall not make delivery of this shipment without payment of freight and all other lawful charges.

_____
(Signature of Consignor)

FREIGHT CHARGES
PREPAID ☐   COLLECT ☐

RECEIVED, subject to the classifications and lawfully filed tariffs in effect on the date of issue of this Bill of Lading, the property described above in apparent good order, except as noted (contents and condition of contents of packages unknown), marked, consigned, and destined as indicated above which said carrier (the word carrier being understood throughout this contract as meaning any person or corporation in possession of the property under the contract) agrees to carry to its usual place of delivery at said destination, if on its route, otherwise to deliver to another carrier on the route to said destination. It is mutually agreed as to each carrier of all or any of, said property over all or any portion of said route to destination and as to each party at any time interested in all or any said property, that every service to be performed hereunder shall be subject to all the bill of lading terms and conditions in the governing classification on the date of shipment. Shipper hereby certifies that he is familiar with all the bill of lading terms and conditions in the governing classification and the said terms and conditions are hereby agreed to by the shipper and accepted for himself and his assigns.

This is to certify that the above-named materials are properly classified, described, packaged, marked and labeled and are in proper condition for transportation according to the applicable regulations of the Department of Transportation.
Per _____

PLACARDS REQUIRED

SPECIAL INSTRUCTIONS: _____

PLACARDS SUPPLIED   ☐ YES ☐ NO — FURNISHED BY CARRIER
DRIVERS SIGNATURE: _____

SHIPPER: _____   CARRIER: _____
PER: _____   PER: _____
DATE: _____   DATE: _____

8 - B L S - C 3   (REV. 5/87)

CONTAINS HAZARDOUS MATERIALS
FOR HELP IN CHEMICAL EMERGENCIES INVOLVING SPILL, LEAK,
FIRE OR EXPOSURE CALL TOLL-FREE 1-800-424-9300 DAY OR NIGHT

Figure 2.6 Straight Bill of Lading. (Reprinted from J.J. Keller and Associates, Neenah, Wisconsin.)

biennial report must also be kept on file for at least three years. The biennial report gives the Board of Equalization the total waste tonnage shipped for tax purposes. Permit fees are also based on total tonnage, which is addressed in the biennial report.

The Price of Noncompliance

Most regulations, such as RCRA, allow the agency to conduct inspections and, if problems are noted, to initiate enforcement. Depending on whether the organization is just a generator or is a TSDF, there are approximately six different types of inspections allowed under RCRA. If problems are noted, the EPA can then initiate one or more of the following, in increasing order of severity: warning letter, notice of violation, compliance order, corrective action order, civil penalty, permit suspension, civil court action or criminal court investigation [Blackman 1993].

Noncompliance can result in a multitude of negative outcomes. Most are hard to quantify and can be devastating to the individual and organization. Of course, the death of humans and other organisms exposed to different hazardous materials and wastes is by far the most severe.

Fines

Fines are probably the easiest measure of noncompliance to define. A sampling of fines is discussed in Chapter 13. Fines can go as high as $250,000/day.

One of the most common EPA fines has been in reference to the land bans. Since these regulations are very complex, it is easy for generators to incorrectly identify a waste or send a prohibited waste to a disposal site.

The dollar amount of fines has increased dramatically over the last few years. New legislation such as California SB 2057, which was passed in 1992, allows for additional penalties for hazardous substance release. SB 2056 also passed in 1992, and allows the governor to collect treble damages from liable parties violating cleanup orders.

The amount of enforcement and level of fines are greatly different from one country to the next. In the Far East, there is very little enforcement of regulations.

AB 2249, the California Criminal Liabilities Act, was effective January 1, 1991. This makes a manager or a corporation civilly liable for damages caused by dangerous products or business practices. Up to three years in prison and $1,000,000 in fines can be assessed.

Imprisonment

Imprisonment is also discussed in Chapter 13. Six years is one of the highest terms and applies to California Prop. 65. In 1991, prison terms reached an all-time high. Figure 2-7 illustrates the change in U.S. jail time due to environmental crimes from 1984 to 1991. In 1984, there were six months of time served; in 1991, there were 903 months of time for all individuals sentenced that year. In recent years, some of the increase shown is due to the addition of some nonofficer, nonowner jail time. Traditionally, prosecutors only went after high-profile individuals. This is not entirely the case any longer, as some management and director-level employees have also been sent to prison.

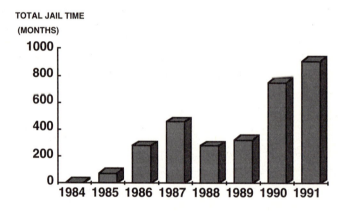

Figure 2.7 Jail Sentences. Total jail time, in months, for all individuals sentenced for environmental crimes for that year. (Source: EPA.)

Cost Reimbursement for Cleanup and Damage

Many companies have had to pay huge sums of money for improperly handling HMW. Cleanup is the responsibility of the generator or an "involved" deep pocket. CERCLA has given the EPA the power to collect millions of dollars from many organizations.

Court Settlements

Settlements for damages and cleanup have occurred both in and out of court. The sky is the limit in these situations. Several million dollars is not an uncommon settlement amount. The Exxon Valdez settlement in March 1991 was reported to have exceeded one billion dollars.

Bad Publicity

Probably as damaging to any organization as all of the preceding is the bad publicity generated with HMW incidents. One incident can lead to the downfall of an organization. This is due to the highly emotional nature of the subject.

Case Studies

Dexter Corp—Hartford, Connecticut

The *San Francisco Chronicle* reported, on September 11, 1992, that one of the biggest fines to date under the Clean Water Act was assessed against the Dexter Corp. This included $9 million in civil penalties and $4 million in criminal penalties for alleged discharge of carbon disulfide from its paper products plant into the Connecticut River through a discharge pipe. According to the EPA, some of the chemical was also spilled when workers turned over partially empty drums.

Wagner Seed

If a spill results because of an act of God, who will be held responsible? In 1986, Wagner Seed Company was held responsible for a lightning bolt causing an accidental spill from its hazardous waste storage facility. The EPA made the ruling under CERCLA Section 106. Wagner Seed Company stated that they should not be considered a responsible party. In an act of God, the court ruled, there is no third party. In this case, there is only the EPA, God and Wagner. Reimbursement from EPA and God is doubtful. That leaves only Wagner [Second Circuit Court].

CHAPTER

THREE

Hazardous Material and Waste Determination and Classification

Perhaps one of the most confusing and difficult tasks that confront a manufacturer and a generator is to determine whether their substance is hazardous and, if so, which class it best fits into. If the letter of the law is taken literally, almost everything, in certain situations, could be considered hazardous. For example, rain water occasionally meets corrosive criteria. Fortunately, many, but not all, regulators allow some realism to be exercised. Once the determination has been made, however, it is difficult, if not impossible, to get a material or waste delisted.

It is up to the manufacturer (in reference to a hazardous material) or the generator (in reference to a hazardous waste) to determine whether its organization has HMW. In general, the first questions to ask are qualitative and include what chemicals are present, in what forms and with what hazardous characteristics. Next, the quantitative question must be asked, or how much is there? If an organization has a chemical that is listed (such as in Appendix A) as a hazardous waste or hazardous material or exhibits the listed characteristics, they must comply with the applicable HMW regulations. Compliance is also required if they have an unlisted chemical with the regulatory characteristics of an HMW. For example, if the chemical exhibits ignitibility, corrosivity, reactivity or toxicity or poses a potential hazard to human health or the environment, it will probably be designated a hazardous waste.

There is some similarity in the way various countries make the HMW determination. In the U.S., lists of both wastes and characteristics are used. Belgium, Denmark, France, Germany, the Netherlands, Sweden and the U.K. also use an "inclusive list" and, in certain cases, also use characteristics or criteria.

Hazardous Material Determination

To make a hazardous material determination, the manufacturer can check lists, do tests, check the literature (epidemiological data) or just assume that the component is hazardous. The reverse is not true, however. One cannot assume a material is nonhazardous. Various hazardous material lists are available, such as the DOT hazardous materials table in 49 CFR Part 172. The State of California Chemical List of Lists is another good resource to check.

Hazardous Waste Determination

Hazardous waste can come in almost any shape or form, making determinations a challenge. Many liquids, semi-solids, solids and sludges that one would assume are not hazardous turn out to be so.

There are some very general considerations that will help in terms of hazardous waste determination. Does the local POTW or landfill prohibit that type of waste? Are there air, water or solid waste regulations that restrict on-site treatment or discharge? If any of the answers are yes, the waste is probably going to be designated hazardous. By asking these common-sense questions, considerable time and money can be saved in the determination process.

Usually, the determination concerning whether a waste is hazardous or not is based on laboratory tests. The determination can be based on existing data if the data are adequate. New laboratory tests are usually carried out to make the determination. The laboratory tests may even show that a listed waste should be delisted. Lab tests can also be used to determine the concentration of a listed waste in a matrix (e.g., soil). If the listed waste exceeds the concentrations outlined in 40 CFR 261, the matrix is considered hazardous.

Overall, determinations are made in accordance with either lists of wastes or characteristics. The EPA concentrates more on listed wastes. Some states, such as California, concentrate more on characteristics.

RCRA (EPA) Hazardous Waste

Figure 3-1 presents the EPA's flow sheet for defining whether a waste is hazardous or not. It also presents some of the regulatory requirements for hazardous waste and several other categories of waste. Figure 3-1 is a general overview type of flow sheet since several of the steps have their own detailed substeps. For example, the entry in the center, "Does the waste exhibit any of the characteristics specified in Part 261, Subpart C," actually has subflow sheets.

Certain wastes are designated RCRA wastes and are regulated by the EPA. These wastes are byproducts of manufacturing processes or discarded commercial products. For a waste to be considered hazardous by EPA, it must meet one of the following conditions:

- Listed in 40 CFR Part 261—40 CFR Part 261 includes three important federal lists of hazardous wastes, along with their hazardous waste numbers. The first list is from 40 CFR Section 261.31 and is Hazardous Wastes from Non-Specific Sources (F series wastes). The second list is from 40 CFR Section 261.32 and is Hazardous Wastes from Specific Sources (K series wastes). The third list is from 40 CFR Section 261.33 and is Discarded Commercial Chemical Products, Off-Specification Species, Container Residues and Spill Residues (P and U series wastes). These three lists are used very frequently by HMW managers. There are also D series wastes, which are waste constituents that fail the toxicity characteristic leaching procedure (TCLP) test.

- Exhibits ignitibility, corrosivity, reactivity and/or toxicity upon analysis. There are specific definitions for each of these characteristics which are presented later.

- Is a byproduct from the treatment of a hazardous waste (unless excluded)

- Is a substance not excluded by RCRA

- Is a mixture containing a listed hazardous waste

State-Listed Hazardous Waste (Non-RCRA)

States may regulate some wastes as hazardous that the EPA may not consider hazardous. For example, in California waste oil is considered hazardous, but not by the EPA. California-regulated hazardous wastes include hazardous, extremely hazardous and special wastes. The lists in Title 22 should first be checked. If the hazardous waste is not listed, the hazardous

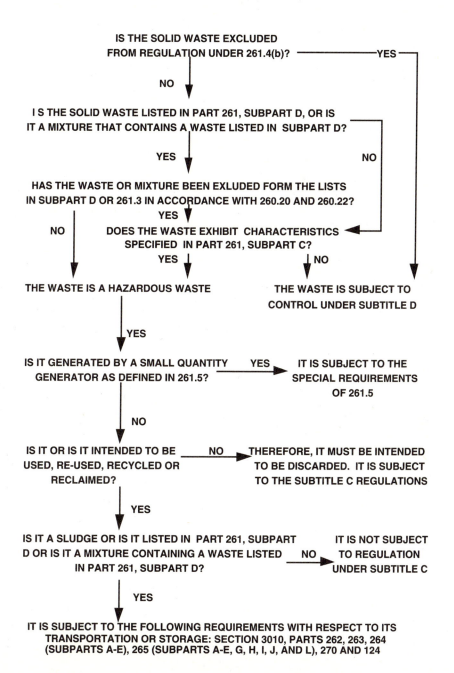

Figure 3.1 Definition of a Hazardous Waste.(Reprinted from EPA, *40 Code of Federal Regulations Parts 300-399,* Washington, 1990.)

characteristics listed in Title 22 must be assessed. For a waste to be considered hazardous by California, it must meet one of the following conditions:

■ Listed in CCR Article 4 and Appendix X of Chapter 11 (either the waste or the waste components).

■ Not excluded under CCR 66261.4 or HSC 25143.2.

■ Exhibits ignitibility (CCR 66702), corrosivity (CCR 66708), reactivity (CCR 66705), and/or toxicity (CCR 66696). The toxicity characteristic is assessed by the Waste Extraction Test (WET). Until 1991, California regulated infectious waste as a hazardous waste. After that date it was regulated as a medical waste under the California Medical Waste Management Act of 1990.

■ Meets the definition (HSC 25117) for hazardous waste: a waste, or combination of wastes, which because of its quantity, concentration, physical or chemical characteristics may either:

• Cause or significantly contribute to an increase in mortality or an increase in serious, irreversible or incapacitating reversible illness.

• Pose a substantial present or potential hazard to human health or environment when improperly treated, stored, transported, disposed or otherwise managed.

Other Agency-Regulated Hazardous Wastes

In addition to the state and EPA, in special situations other agencies may make a determination of whether a substance is an HMW. For example, the federal DOT may make determinations. A generator can also ask the transporter or disposal facility for HMW determination assistance. Last, some local agencies, such as the California RWQCB, may also make some hazardous waste determinations.

Hazardous Material and Waste Classification Systems

Once the determination has been made that a substance is an HMW, it falls into one or more HMW categories or classification systems. There are at least four or more classification systems for HMW. Fortunately, there are many similarities. Table 3-1 compares the four most common classification systems. These classification systems have their own verbal descriptions, number codes, color codes and symbols [Allegri 1986]. The DOT/United Nations system and the OSHA system deal with HMW. The California and federal EPA systems deal only with hazardous waste.

Table 3.1 Major Classes of Hazardous Substances

DOT/United Nation Standard/IMCO Hazard Classes—Used primarily on labels, placards, bills of lading and manifests for the shipment of hazardous materials and waste:

1. Explosives
2. Gases - flammable, non-flammable and poisonous gases
3. Flammable and combustible liquids
4. Flammable solids (spontaneously combustible and dangerous when wet materials)
5. Oxidizers and organic peroxides
6. Poisonous and harmful
7. Radioactive substances
8. Corrosive materials
9. Miscellaneous dangerous material

CAL EPA Hazard Classes—Used primarily in hazardous waste class determination:

1. Ignitibility
2. Toxicity
3. Corrositivity
4. Reactivity
5. Extremely hazardous

EPA Hazard Classes—Used primarily in hazardous waste class determination:

1. Ignitible
2. Toxic
3. Corrosive
4. Reactive
5. Acute Hazardous

OSHA Hazard Classes:

1. Physical Hazards Group
   a. Combustible liquid
   b. Compressed gas
   c. Explosive
   d. Organic peroxide
   e. Pyrophoric
   f. Unstable/reactive
   g. Water reactive
   h. Flammable aerosol
   i. Flammable gas
   j. Flammable liquid
   k. Flammable solid
   l. Oxidizer

2. Health Hazards Group
   a. Carcinogen
   b. Corrosive
   c. Highly toxic
   d. Irritant
   e. Toxic
   f. Hepatotoxin
   g. Nephrotoxin
   h. Neurotoxin
   i. Blood toxin
   j. Lung toxin
   k. Reproductive toxin
   l. Cutaneous hazard
   m. Eye hazard

DOT/United Nations/IMCO System of Hazard Classes

On January 1, 1991, the new DOT/United Nations Performance Oriented Packaging Standards were incorporated into the DOT regulations. Many changes occurred in terms of selecting packaging, shipping documents, labeling, marking, and placarding. The DOT/United Nations system is probably the most important classification system and must be followed to the letter when preparing manifests, posting labels and placards. The system concentrates on physical, chemical and major hazardous characteristics. Not considered in terms of class placement is the variation within a class from slight to severe degree of hazard in an emergency. A chemical may be placed in several of these categories. Details can be found in 49 CFR. The classes include:

Class 1—Explosives and New Explosives Explosives are any chemical compound the purpose of which is to function by explosion and include detonating and combustion explosives, small arms ammunition, fireworks and blasting agents. Explosives must be protected from heat, shock and contamination. Under the new classification effective January 1, 1991, explosives are reclassified into subclasses 1.1 through 1.6.

Class 2—Flammable, Nonflammable and Poisonous Gases This class includes gases which are under normal temperature (68 degrees F) and atmospheric pressure (14.7 psi). It includes both flammable (hydrogen, propane, etc.) and nonflammable gases (nitrogen, oxygen, etc.). There are many other hazards of compressed gases, including explosion, asphyxiation, reactivity, ground movement, frostbite and toxicity.

Class 2 materials are subdivided into three classes. 2.1 includes flammable gas, 2.2 is for nonflammable, nonpoisonous compressed gas and 2.3 is for gas that becomes poisonous when inhaled.

Class 3—Flammable and Combustible Liquids This is one of the most common hazardous material classes and includes gasoline and solvents. It includes flammable liquids that have a flash point below 100 degrees F and combustible liquids that have a flash point between 100 and 200 degrees F. Flash point is the minimum temperature at which a liquid gives off vapor within a test vessel in sufficient concentration to form an ignitible mixture with air. Certain flammable liquids are also pyrophoric, or can self-ignite. This class has many of the same multiple hazards as mentioned above for compressed gases.

Class 4—Flammable Solids, Spontaneously Combustible and Dangerous When Wet Materials Flammable solids burn with speed greater than combustible substances. These can also be readily ignited. There are a wide variety of characteristics in this class. For example, some flammable solids spontaneously combust (cotton waste). Other flammable solids are water-reactive (metallic sodium) or air-reactive (white phosphorus). Many flammable solids are toxic or corrosive before and/or after contact with air or water.

There are three subclasses for class 4 substances. 4.1 is for flammable solids, 4.2 is for spontaneously combustible materials and 4.3 is for materials that are dangerous when wet.

Class 5—Oxidizers and Organic Peroxides Oxidizers are substances that contain significant amounts of chemically bonded oxygen. They yield oxygen readily and therefore stimulate combustion of other substances. Examples of oxidizers include sodium chlorate, potassium and nitric acid (used in fertilizers). Oxidizers should not be stored with combustible materials. Some of the multiple hazards of oxidizers include reactivity, ignitibility, combustibility and explosivity.

There are two subclasses for class 5. 5.1 is for oxidizers and 5.2 is for organic peroxide. 5.2 is further subdivided into seven classes based on the detonation characteristics of the contained material.

Organic peroxides are used in many industrial applications, especially in plastic manufacturing (adhesive component). They are unstable and can become explosive. Examples of organic peroxides include benzoyl peroxide, lauroyl peroxide and peracetic acid. Multiple hazards include flammability and the release of heat and toxic byproducts.

Class 6—Poisonous and Infectious Substances Poisons may enter the body via ingestion, injection, absorption or inhalation. Class 6 is subdivided into 6.1 and 6.2. 6.1 is for poisonous material based on oral, dermal and inhalation toxicities indicated by lethal dose 50 or lethal concentration 50. 6.1 also includes irritating materials. 6.2 is for infectious substances.

Most infectious substances are etiologic agents or living microorganisms. The microorganisms may enter the body via inhalation, ingestion or absorption and cause disease. The most common example would be the biological and viral cultures and samples associated with medical clinics and biological research facilities, such as the measles virus. An exposure to etiologic agents will result in conditions very similar to those caused by poisons. In general, infectious substance includes diagnostic specimens, biological products and regulated medical waste.

Class 7—Radioactive Materials Radioactive materials release alpha, beta and/or gamma radiation and are subdivided into three classes. These materials are commonly found in nuclear power, medicine, industry, and research. Common examples include radioactive iodine and cobalt, used in medicine, uranium, used in power generation, and tritium, used in biological research. Multiple hazards include tissue damage, mutation, contamination of the environment, and heat generation.

Class 8—Corrosive Materials Corrosive materials are liquids or solids that can destroy human tissue by skin contact or inhalation or corrode metals (at a rate of 0.246 inch a year on steel or aluminum at 131 degree F). Some corrosives are also oxidizers, toxic and unstable, and decompose when heated. Acids are the most common, such as sulfuric acid and hydrochloric acid. Sodium hydroxide (lye) and potassium hydroxide (potash) are also common corrosives. Multiple hazards include corrosion, possible support of combustion, rapid heat release, toxicity, and instability.

Class 9—Miscellaneous Hazardous Materials Class 9 includes materials that present a hazard during transportation but do not meet the definition of any other hazard class. They are materials that have an aesthetic impact or noxious property that causes annoyance or discomfort.

Other Regulated Materials (ORMs) used to be part of Class 9; however, ORMs have been separated from Class 9 materials under the new DOT rule amendments. Under the new definition, ORMs mean a material, such as consumer commodity, which, although otherwise subject to the applicable DOT regulations, presents a limited hazard during transportation due to its form, quantity and packaging.

State Classification System

As an example of a state hazardous waste classification system, California will be given. The California EPA classes are very similar to the federal EPA classes. One of the biggest differences, however, is in the toxicity class and the tests involved to determine toxicity.

Ignitibility Wastes are classified as ignitible if they meet one of four criteria. A liquid that has a flash point less than 60 degrees C or 140 degrees F is ignitible. A nonliquid capable of causing fire through friction or absorption of moisture is ignitible. Flammable compressed gas or an oxidizer are also considered ignitible.

Toxicity   A waste is considered toxic if it meets any one of the following criteria: has an acute oral LD50 < 5000 mg/kg, has an acute dermal LD50 < 4300 mg/kg, has an acute inhalation LC50 < 10,000 ppm as a gas or vapor, has an acute aquatic 96 hr LC50 < 500 mg/l, contains a listed substance ≥ 0.001% by weight or is listed in 40 CFR 261. The main toxicity test used by California is the Waste Extraction Test (WET). The WET test uses citric acid for 48 hours for 19 metals and 18 organic compounds.

Corrosivity   To meet the corrosivity criteria, a waste must meet one of the following two conditions: An aqueous liquid that has a pH ≤ 2 or ≥ 12.5 or corrodes steel (SAE 1020) at a rate greater than 6.35 mm/year at 55 degrees C.

Reactivity   Reactive wastes are wastes that meet one of the following conditions: normally unstable, reacts violently with water, forms explosive mixtures with water, generates toxic gases, is a cyanide- or sulfide-bearing waste, capable of detonation or explosive reaction or is an explosive according to 49 CFR 173.

Extremely Hazardous   There are various conditions, one of which must be satisfied for a waste to be termed extremely hazardous: acute oral LD50 ≤ 50 mg/kg, acute dermal LD50 ≤ 43 mg/kg, acute inhalation LC50 ≤ 100 ppm as a gas or vapor, contains a carcinogenic substance listed in Section 66261.24 of 22 CCR at a concentration ≤ 0.1% by weight, has been shown through experience or testing to pose an extreme hazard, is water-reactive, or contains persistent or bioaccumulative substance equal to or larger than the listed TTLC.

Federal EPA Hazard Classes

Federal EPA hazard classes can be found in 40 CFR 261 subpart C. An overview of the federal hazardous waste classes is shown in Figure 3-2.

Ignitible Waste (Hazard Code I)   An ignitible waste is a liquid with a flash point less than 60 degrees C, a solid capable of causing fire that burns so vigorously and persistently that it creates a hazard, an ignitible compressed gas or an oxidizer. An example is used solvent.

Corrosive Waste (Hazard Code C)   A corrosive waste is a liquid that has a pH ≤ 2 or ≥ 12.5 or corrodes steel at a rate greater than

Ignitible

Wastes that are combustible or flammable, such as:

- *Paint Wastes*
- *Degreasers*
- *Solvents*

Corrosive

Wastes that dissolve metals, other materials, or burn the skin, such as:

- *Waste Rust Removers*
- *Alkaline Cleaning Fluids*
- *Waste Battery Acid*

Reactive

Wastes that are unstable or undergo rapid or violent chemical reaction with water or other materials, such as:

- *Cyanide Plating Wastes*
- *Waste Bleaches*
- *Other Waste Oxidizers*

Toxic

Wastes that are tested and found to contain high concentrations of heavy metals or chemicals that could pose a threat to public health or the environment, such as:

- *Mercury*
- *Cadmium*
- *Lead*
- *Specific Pesticides*

Figure 3.2 EPA Hazardous Classes. (Reprinted from EPA, 1993.)

6.35 mm/year at a temperature of 55 degrees C. Examples include used pickle liquor from steel manufacturing.

Reactive Waste (Hazard Code R)   A waste is categorized as reactive if it exhibits any of the following characteristics: normally unstable and undergoes violent change without detonating, reacts violently with water, forms explosive mixtures with water, generates toxic gases, vapors or fumes, is a cyanide- or sulfide-bearing waste that when exposed to pH

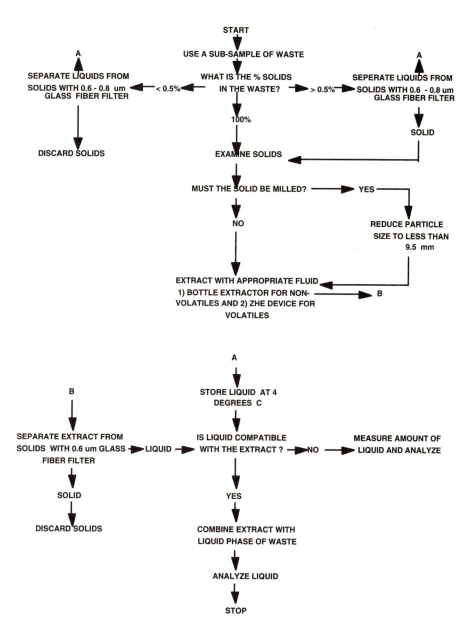

Figure 3.3 Toxicity Characteristic Leachate Procedure. (Reprinted from EPA, *40 Code of Federal Regulations,* Ch. 1, Washington, 1990.)

between 2 and 12.5 generates gases, vapors or fumes, is capable of detonation or explosive reaction or is a forbidden explosive.

Toxic Waste (Hazard Code T)   Part of the challenge in determining toxicity is in the use of a standard method that everyone can understand and use consistently to prepare and analyze their waste. The EP Toxicity Test was developed a few years back by the EPA to accomplish this goal. Because of problems, however, in September 1990 the EP Toxicity Test was replaced with the Toxicity Characteristic Leachate Procedure (TCLP) test. Figure 3-3 is a flow diagram that presents the major steps to be taken when the TCLP is utilized. If, after following this procedure, the extract exceeds published levels, the substance is determined to be toxic and is therefore a hazardous waste. The TCLP is now used to determine toxicity at the federal level. A substance is considered a toxic hazardous waste if a sample of it is processed as illustrated in Figure 3-3 and its extract contains any of the contaminants listed in 40 CFR 261.24 at or exceeding the specified levels. Examples of the listed substances include arsenic, barium, cadmium, chromium, lead, mercury, selenium, silver, endrin, lindane, methoxychlor, and toxaphene.

Acute Hazardous Waste (Hazard Code H)   Acute hazardous waste is a special class of waste that is more highly toxic and includes discarded commercial chemical products, containers, spill residues, manufacturing chemical intermediates, off-specification commercial chemical products and manufacturing chemical intermediates. These wastes are commonly referred to as extremely hazardous wastes.

This waste determination is important in establishing the reportable quantity (RQ) for emergency response and reporting requirements. The RQs for acutely hazardous wastes are often low, commonly less than 50 pounds.

OSHA Hazard Classes

There are 25 OSHA hazard classes arranged into two major groups, physical hazards and health hazards. The physical hazards classes include combustible liquid, compressed gas, explosive, organic peroxide, oxidizer, pyrophoric, unstable, water-reactive, flammable aerosol, flammable gas, flammable liquid and flammable solid. The health hazard classes include carcinogen, corrosive, highly toxic, irritant, toxic, hepatotoxin, nephrotoxin, neurotoxin, blood toxin, lung toxin, reproductive toxin, cutaneous hazard and eye hazard.

Case Study

Life Sciences—Hopewell, Virginia

The following case study illustrates how an incident occurred partially because of lack of hazardous classification. As a result, new regulations were implemented. Among other things, these regulations significantly strengthened the classification of hazardous substances.

In 1973, Allied Chemical subcontracted the production of a pesticide, Kepone, to Life Sciences Products. Life Sciences was started by two former employees who operated out of a converted gas station near the James River. Within 16 months, health problems among its employees surfaced, prompting investigations by OSHA, EPA and the state of Virginia. Kepone dust was found on the floor several inches deep, and employees complained of vapors that irritated their eyes and skin. Health and safety regulation violations were uncovered, and illegal discharge of Kepone into the James River was identified.

During a 16-month period, several complaints were made with little or no action. Finally, in July 1975, production was shut down and Life Science fined for willfully failing to use feasible engineering control measures to prevent harmful levels of exposure and willfully not providing or maintaining personal protective equipment.

The problems in this case developed from multiple causes. Regulatory agencies failed to identify Kepone as a hazard and to act on individual complaints. The need for regulating the production of chemicals and associated waste disposal was identified, resulting in the promulgation of regulations that classify waste products and the regulation of manufacturing, reporting and disposal operations [Wentz 1989].

FOUR

Generators of Hazardous Wastes

Introduction

There are many more categories of generators of HW than most people would think. In addition to the obvious heavy industrial sources are the biological waste generators, households, transporters, farmers, military and miscellaneous organizations. This includes just about everyone.

Most of the principles discussed in this book would apply to all of the different generators mentioned above. Certain generators, such as households and farmers, follow very few hazardous waste principles, guidelines or regulations. This problem is compounded by the fact that agencies have concentrated almost totally on the other types of generators. Ignorance or indifference is not uncommon in some generator populations.

Generators can also be divided into conditionally exempt, small-quantity generators (SQGs) and large-quantity generators (LQGs). Conditionally exempt generators produce less than 100 kg/month of hazardous waste or less than 1 kg/month of acutely hazardous waste. A SQG is one who produces more than 100 kg/month and less than 1000 kg of hazardous waste/month or not more than 1 kg of acutely hazardous waste/month. LQGs are those that produce more than 1000 kg of hazardous waste/month or more than 1 kg/month of acutely hazardous waste. Both SQGs and LQGs must obtain an EPA ID number, manifest, keep records and report.

Heavy Industry

The majority of the discussions that have occurred or will occur in this book apply most directly to industry and utilities. Billions of pounds of hazardous waste are generated by industry each year. Table 4-1 presents examples of hazardous waste generated from different industrial sources. There are hundreds of other types of industrial generators that are not shown; however, the predominant waste types are shown in Table 4-1

Table 4.1 Examples of Hazardous Waste
Generated by Businesses and Industries

Waste Generators	Waste Type
Chemical Manufacturers	Strong acids and bases Spent solvents Reactive wastes
Vehicle Maintenance Shops	Heavy metal paint wastes Ignitible wastes Used lead acid batteries Spent solvents
Printing Industry	Heavy metal solutions Waste inks Spent solvents Spent electroplating wastes Ink sludges containing heavy metals
Leather Products Manufacturing	Waste toulene and benzene
Paper Industry	Paint wastes containing heavy metals Ignitible solvents Strong acids and bases
Construction Industry	Ignitible paint wastes Spent solvents Strong acids and bases
Cleaning Agents and Cosmetics Manufacturing	Heavy metal dusts Ignitible wastes Flammable solvents Strong acids and bases
Furniture and Wood Manufacturing and Refinishing	Ignitible wastes Spent solvents
Metal Manufacturing	Paint wastes containing heavy metals Strong acids and bases Cyanide wastes Sludges containing heavy metals

Reprinted from EPA, *Solving the Hazardous Waste Problem/EPA's RCRA Program*, 530-SW-86-037, Washington, 1986.

(strong acids, bases, spent solvents, reactive wastes, heavy metal wastes and ignitible wastes). Industrial hazardous wastes have been the center of attention for many years. Industry is controlled by many laws, regulations and agencies. RCRA and TSCA are two important examples.

Examples of hazardous wastes that are commonly generated in one type of industry, electronics manufacturing, will be given. These include waste acids (phosphoric, hydrochloric, nitric, sulfuric), waste solvents (acetone, alcohol, freon, trichloroethane, xylene), waste coolants (ethylene glycol), heavy metal sludges (nickel, zinc, chromium) and a wide variety of other wastes (oil, ammonia, sodium hydroxide, etc.).

Biological Waste Generators and Laboratories

Biological wastes were not paid much attention until the AIDS epidemic surfaced and medical wastes started washing up on East Coast shores in considerable amounts. Now significant concern is directed toward most biological waste, and employees are putting almost everything into red bags for incineration. This has caused the incinerators to exceed their capacity. Laboratories, hospitals and biotechnology generate this special type of hazardous waste.

There are four major types of waste in this category: infectious, injurious, noninfectious and hazardous. Infectious waste is the most troublesome because it may transmit disease. It includes tissue or cell cultures, blood, and body parts. Injurious waste, such as needles, may inflict wounds. Noninfectious waste includes packaging materials. Hazardous wastes include chemotherapy drugs, narcotics and many outdated or unused chemicals, found in most labs.

Handling procedures for these wastes vary greatly depending on the waste. Generally, radioactive waste is kept in lead-lined containers, picked up regularly and disposed of offsite. Hospital infectious wastes and chemotherapy chemicals are placed in double red, plastic-lined containers, picked up by gloved housekeeping personnel and then incinerated. The hazardous chemicals are neutralized or transported to a treatment, storage or disposal facility (TSDF).

Infectious waste generated during biological research, genetic engineering or other activities in biotechnology may require extreme care in handling. The infectious agent may be concentrated many times above normal, and the infection process may not be clearly understood. The extra care should include laminar flow hoods, negative pressure rooms, immediate incineration and respirators. Common problems noted with these generators include improper hoods or hoods that are not properly vented;

disposal facilities that are sometimes inadequate and may involve use of the sewer line; and eating and drinking in the laboratory.

Illegal drug laboratories could also be considered in this class of generators. During 1989 and early 1990, about one illegal drug lab/month was discovered in Santa Clara County in California. For obvious reasons, the labs do not properly dispose of waste, and most of the labs are set up in rented homes or leased warehouses. The wastes are commonly found abandoned at the building, and it is believed that a large quantity of waste is illegally disposed of in creeks, on the ground and in class III landfills. Hazardous wastes of this type commonly include purification solvents and improperly mixed drug batches. The primary hazards presented by these wastes include flammability, toxicity and explosivity.

Households

The public generally feels that households don't contribute much hazardous waste and, if so, it is of low toxicity. This is wrong on both counts. In addition, household hazardous waste is essentially not regulated by agencies. Approximately 19% of household trash is hazardous waste. This makes it a very large source.

A city the size of San Jose would produce, each month, approximately two tons of bowl cleaner, 96 tons of liquid cleaner, 25 tons of used motor oil, and many tons of pesticides, herbicides and miscellaneous chemicals. This is more than most medium-sized industrial generators.

A large volume of household hazardous wastes go into sanitary landfills (dump or Class III landfills) by accident, ignorance or lack of reasonable options, with the weekly trash pickups. Class III sites are not designed to handle hazardous wastes. This results in environmental pollution. Examples of these common household hazardous wastes that incorrectly end up in Class III landfills include certain:

- Household cleaners

- Paint products—solvents, wood preservatives, turpentines, oil-based paints, paint remover

- Pesticides and herbicides—fungicides, moth balls, insect sprays

- Automotive products—gasoline, antifreeze, used motor oil, batteries

- Miscellaneous—outdated photo supplies, medicines, glues, furniture polish

Municipal garbage is exempt from regulation as RCRA hazardous waste. Ash from municipal garbage does contain concentrations of leachable toxics that exceed RCRA criteria and would be considered hazardous from other sources. For example, heavy metals are commonly found in incinerator ash. The U.S. District Court in New York decided on November 22, 1989, that municipal garbage byproducts, such as ash, were excluded from RCRA.

One very easy way for the public to help manage hazardous wastes is to read labels on all products before buying them and before disposal. If possible, a product without a hazardous substance should be purchased. The public must also educate and regulate themselves. For this to occur, there needs to be a change in personal habits and attitudes.

Household hazardous waste cleanup programs are starting to be set up in many communities. These are sponsored by teams involving agencies (city, county, state, federal), industry and special citizen groups. The programs usually include advertising, collection, analysis, labeling, transporting and disposing of citizens' hazardous wastes from a common collection point. For example, there is a transfer/collection point in South San Francisco for citizens to use five days per week.

Transporters

HMW transporters are also generators of hazardous wastes in certain situations. If they spill a load of hazardous materials, they become the generators. When their tank trucks have to be cleaned, the cleaning solutions may be considered hazardous. Sludges that accumulate in tank trucks would also be considered hazardous waste.

Farmers

Many agricultural activities employ the use of pesticides, fertilizers, rodenticides and other chemicals. Outdated, banned, unused or spilled agricultural chemicals become a hazardous waste problem for farmers. Many of these problems have been and still continue to be ignored.

On September 21, 1990, AB 563 (Hannigan) became law in California (California Health and Safety Code 25207-25207.13). This bill relates to hazardous waste disposal for farmers. The bill authorizes counties to develop a collection and transportation program for hazardous wastes from farmers.

Military

Military hazardous waste generators could be considered in a separate class because of several reasons. Certain aspects of the regulations apply differently to military organizations. Military waste cleanup lags behind civilian cleanup in many cases. Commonly military hazardous waste tends to be more mixed than civilian hazardous waste. The increase in military base closures will probably result in significant new quantities of hazardous waste generated, especially asbestos, cleaners, explosives and other wastes.

Much of the military waste is also high-level radioactive mixed waste. Between the military and utilities, radioactive waste presents a real challenge in terms of proper disposal. Radioactive and chemical waste cleanup at U.S. nuclear weapons and military installations is estimated to cost about $130 billion, according to the March 27, 1989, *U.S. News and World Report*.

Other Miscellaneous Organizations

Agencies, small businesses and most other organizations generate hazardous waste. Most organizations will generate the following unused, outdated or spilled chemicals or byproducts which would be considered hazardous waste: copy-machine toner, oil-based paints, pesticides, cleaners, and automotive products.

Case Study

Disneyland

No free rides for Disney. A $550,000 settlement was reached in a toxic waste complaint against Disneyland. This was in reference to disposal of acetone, paint thinner and cleaning materials from the Anaheim amusement park. For approximately two years, the contractor, Kens Oil Company, hauled 14,000 gallons of material to Wyoming and Utah disposal facilities that did not have the proper permits, according to reports. Disney officials claimed they were unaware of the illegal disposal, which resulted in 38 violations [*Los Angeles Times* 1990].

FIVE

Examples of Hazardous Materials and Waste

Introduction

There are thousands of hazardous materials and wastes. Some of these are listed in 40 CFR 261.31, 261.32, 261.33(e), 261.33(f), 302 (see Appendix A) and in California 22 CCR Article 9. In this chapter, only a few of the more common ones will be mentioned, including PCBs, asbestos, sulfuric acid, lead, ethylene glycol ethers and a few hazardous air pollutants.

Polychlorinated Biphenyls (PCBs)

General Information

PCBs are a group of chlorinated hydrocarbons with a specific gravity of 1.17. Because of this, they will sink in water (one gross method of identification). They range in appearance from a heavy oily liquid to a waxy solid [Crouth 1989].

Use of PCBs

PCBs were manufactured in the U.S. from 1929 to 1977. There are considerable quantities of PCBs still in place or available in stored inventories in the United States. PCBs have been used in electrical transformers, switches, voltage regulators, capacitors, hydraulic fluids, carbonless copy paper, paints, adhesives, caulking compounds, road coverings and other products.

Problems with PCBs

There are many problems with PCBs. When burned, PCBs form dioxin, which is very toxic. PCBs are very stable and will remain in the environment. They have been found in the U.S. in soil, water, milk and many types of tissue (including human). They can enter the organism through many avenues, including ingestion, inhalation and absorption. PCBs are toxic at low concentrations and can cause birth defects, cancer and many other diseases. Recent studies, however, have suggested that the risk is much less than originally believed.

Government Regulation of PCBs

There is considerable regulation of PCBs. For example, TSCA prohibits the manufacture and use of PCBs except for totally enclosed use. TSCA also requires labeling and safe disposal. The FDA and EPA have other PCB regulations that address prohibitions, safeguards and final disposal (e.g., by incineration and hazardous waste landfills) [Allegri 1986].

Asbestos

General Information

Asbestos fibers range in length from 0.1 to 10 μ. The fibers can be either friable (and create hazards) or nonfriable. Friable asbestos is likely to release or shed fibers and can be crumbled with hand pressure. Friable asbestos fibers can be generated during mining, milling, manufacture, fabrication, demolition and many forms of abrasion [Crouth 1989].

History and Use of Asbestos

Over 3000 different products have been manufactured with asbestos. This occurred primarily from the 1940s through 1979, with peak U.S. consumption in the early 1970s. It was banned in homes as of 1979. It is still used in brake linings and certain floor tiles. The products with the highest percentages of asbestos include some textiles with 90% asbestos.

Many products contain asbestos which remains nonfriable unless they are sanded, drilled, sawed or abraded in some other way. These include brake linings, plastics, vinyl, asphalt, linoleum, cement pipe and roofing materials.

The products that contain asbestos likely to deteriorate into a friable state deserve the greatest attention. These include pipe covering paper, insulation, fire-retardant cloth, sprayed and troweled insulation coatings (on beams, ceilings, walls, hot water tanks, boilers) and insulating board.

Asbestos-Caused Diseases

Many individuals believe there is no known safe level of exposure for asbestos. Since it may take up to 20 years for some symptoms to appear, the diagnosis and treatment of asbestos-caused diseases is usually delayed and therefore becomes complicated.

There are three major asbestos-caused diseases: asbestosis, lung cancer and mesothelioma. With asbestosis, there is scarring of lung tissue. Asbestos-induced lung cancer involves dysfunction of lung cells. Mesothelioma is a cancer of the lining of the chest. There are also other cancers that asbestos is suspected of causing.

Regulatory Responsibility

Many acts and regulations pertain to asbestos. The National Emission Standards for Hazardous Air Pollutants include asbestos. TSCA covers asbestos in schools. OSHA pertains to worker protection concerning asbestos and limits PEL to 0.2 fibers/cc/8 hrs. This is a time-weighted average. RCRA covers operation of disposal sites handling asbestos. The Clean Water Act sets asbestos levels in effluents. The Mine Safety Health Administration sets standards for the mining of asbestos. The action level is 0.1 fiber/cc. Concentrations at or above this would require worker protection to prevent exposing the workers to concentrations greater than the PEL.

Owners of buildings built prior to 1979 are required to provide notice to occupants of any suspected asbestos-containing construction materials (ACMs) that may be present in the building. This notice is designed to identify those areas of the building in which the occupants should avoid disturbing the materials.

Control

Asbestos control options can vary dramatically depending on the situation. Control usually involves either removal, encapsulation, enclosure or no action. Initially, many people removed asbestos when it was discovered. Recent trends in asbestos management have shifted away from removal and towards management in place. Worker controls or protection can involve various methods. For example, the use of respirators, medical monitoring and air quality monitoring will minimize exposures.

Transporting Asbestos Wastes

Prior to transportation, asbestos should be properly wetted, double-bagged and labeled. A chain-of-custody form and manifest must be completed. Transportation should be in an enclosed vehicle.

Disposing of Asbestos

Asbestos should be disposed of in a land-fill permitted for asbestos wastes (Class II). It should be covered with at least six inches of soil within 24 hours. Final closure should include at least 30 inches more of soil cover. A considerable volume of asbestos ends up in the Class III sanitary landfill, which is not correct. In certain situations, a Class III landfill may even be able to upgrade their systems and obtain approval from the regulators to accept certain types of asbestos.

Pesticides

Pesticides are a big problem since over two billion pounds of pesticides are used each year. Some present an immediate danger to the user, move up through the food chain and may persist in the environment.

Regulation

Pesticides are covered primarily by two acts, both of which seek to control use while allowing benefits. The Federal Insecticide, Fungicide and Rodenticide Act requires all pesticides to be registered by the EPA and data provided to users by the manufacturer. The Federal Food, Drug and Cosmetic Act—Pesticide Amendment establishes tolerances on the amount of residue that may safely remain on food or crops. The tolerance levels specified for pesticides in many regulations are usually 100 times below the level that might harm people or the environment.

Enforcement of the regulations and laws is by the states or EPA. If they find a pesticide that is unsafe, they may issue an informal notice, formal notice, or may seize goods or initiate civil or criminal proceedings, depending on the seriousness of the violation.

Classification and Labeling

There are two general types of pesticide classifications based on use. General use is by anyone according to label instructions. Restricted use is only by trained and certified applicators.

Pesticide labels must be accurate and according to the regulations. Labels must include EPA number, directions for use, precautions and first aid, storage and disposal instructions.

Examples

One of the most prevalent pesticides is dioxin. The most common of the group is 2,3,7,8 tetrachlorodibenzo-p-dioxin. This is a byproduct from the manufacture of several commercial products, such as the pesticide 2,4,5 trichlorophenol. It is one of the most toxic man-made chemicals known and is thought to cause cancer.

Another example is methyl isocyanate, which killed many individuals at Bhopal. It is used in the manufacture of pesticides such as Temik and Sevin. In terms of action, it attacks any part of the body that is moist. Fortunately, it does not remain in the environment [Allegri 1986].

Sulfuric Acid

Sulfuric acid is a common HM used in many processes and equipment. For example, many large industrial batteries contain sulfuric acid.

When the sulfuric acid picks up too many process contaminants or the acid-containing equipment is no longer needed, it becomes a hazardous waste. The waste acid is usually recycled or neutralized. If there is any market for fresh acid or storage capacity, the acid is generally recycled. Otherwise, neutralization is carried out. In either case, a heavy-metal-laden sludge is generated, which must be disposed of in a Class I site. If the sludge is primarily lead, such as from lead acid batteries, lead reclaimers may be able to take it.

Lead

Many products contain lead, such as the batteries mentioned above and lead sheath cable. When these hazardous materials are being used, care must be exercised to minimize handling and the generation of lead dust. Wetting agents are used (e.g., during the reeling of lead cable) to minimize lead dust generation. Gloves and respirators are common measures of protection since generation of dust cannot be completely avoided.

When most lead-containing products are no longer useful, they are usually purchased by lead reclaimers for recycling. In some states, such as California, it is possible to get variances that allow the handling of lead waste as a hazardous material and not a hazardous waste.

Ethylene Glycol Ethers

Ethylene glycol ethers are found in photoresist products and are used to pattern circuits onto thin film wafers. A study done by IBM and the Silicon Industry Association suggests that this chemical has the potential to cause adverse reproductive effects in male and female workers. In January 1993, the EPA initiated an expedited review, and in March 1993, OSHA proposed a PEL of 0.5 ppm.

Management of this hazardous material involves introducing better procedures to minimize exposure and replacement with other products. Exposure reduction is most effective via engineering controls. One administrative control that many exercise is to transfer pregnant employees to other jobs. Ethylene glycol ethers should not be confused with the larger class of glycol ethers, many of which are not considered HMW and are commonly used in industrial and home applications.

Hazardous Air Pollutants

Several toxic air contaminants or hazardous air pollutants are regulated by various agencies, such as the California Air Resource Board (ARB). Authority is provided under Health and Safety Code 39500. Air toxics are also covered under other statutes such as Prop. 65 (Health and Safety Code 25249.5).

The ARB has determined that the following are considered toxic air contaminants in California: asbestos, cadmium, ethylene dichloride, benzene, chromium VI, dioxins, ethylene oxide, ethylene dibromide, carbon tetrachloride, methylene chloride, furans, trichloroethylene, chloroform, vinyl chloride, inorganic arsenic, nickel, perchloroethylene, formaldehyde and 1,3-Butadene.

Vehicles are one example of a hazardous air pollution source. In September 1990, the ARB approved one of the most sweeping vehicle emissions rules in the world. Highlights of the new emissions rules passed in

September 1990 require that auto makers start selling vehicles with cleaner emission characteristics beginning in 1994 in the Los Angeles basin and, in 1997, statewide. By the year 2003, all cars sold in the state must emit at least 70% fewer hydrocarbons and other smog-forming chemicals than the 1993 models. Two percent of all cars sold must produce zero emissions beginning in 1998, rising to 10% by 2003. Electric cars are the only vehicles that presently meet that standard.

The federal government also regulates air pollutants. EPA has designated the following as air toxics: asbestos, benzene, beryllium, coke oven emissions, inorganic arsenic, mercury, radionucleides and vinyl chloride. The National Emission Standards for Hazardous Air Pollutants (NESHAPS) has documented seven hazardous air pollutants.

Case Studies

General Electric—Oakland, California

Leaking tanks, trench burial, spills from a mobile filtering unit and discharges from a lab sink that emptied into the ground caused PCB contamination of soil and ground water at a 24-acre General Electric site in Oakland, California. The contamination occurred from the 1930s to 1975 and was from an insulating fluid, Pyranol, which contained PCBs and trichlorobenzene.

A complaint was filed by a GE employee, and on July 20, 1979, a state inspector visited the site. The inspector found no indication of PCBs being mishandled. The GE employee was not satisfied with the results and requested that another inspection be conducted. Samples were taken, and the contamination was found.

The state directed GE, in November 1979, to remove PCB-contaminated soil for disposal at a Class I landfill. At that time, it was estimated that 20,000 gallons of Pyranol were beneath the site. During the Phase II investigation, PCB levels as high as 5–11,000 ppm in the soil and 0.63–15 ppb in groundwater were found.

A Cleanup and Abatement Order was issued in December 1980. The consultant retained by GE found a much larger volume of contamination than originally anticipated. The original state order was for excavation of contaminated soils. Due to the additional volume, an alternative correction plan was approved by the state and implemented. It consisted of the installation of a

French drain and treatment system and surface sealing with runoff control. The site cleanup activities were accomplished in 1981 [Wentz 1989].

Duluth, Minnesota

A derailed freight train near Duluth, Minnesota, spilled benzene into a river and forced widespread evacuations. The evacuation affected about 2000 in downtown Duluth and low-lying areas of Duluth. A police officer stated, "If you can smell it, you should seek higher ground." Actually, what he should have said was, "If you can smell it, you should already be on higher ground" [*San Jose Mercury News* 1992].

SIX

Sampling and Analysis

Introduction

Sampling and analysis is done to better understand the situation, whether for academic interest or practical application. It is also done to help define or revise standards. In general, it helps one to understand the type and level of impact of HW on the environment or the impact of noncompliance on the operation.

The impact and fate of hazardous wastes on the environment are determined by several interacting variables. Of great importance is water solubility. Also important are soil adsorption, moisture, pH, porosity, composition, surface area, vaporization, degradation, chemodynamics (structure, weight, polarity), bioaccumulation, medium characteristics (pH, flow, other con-taminants) and microbial conversions of toxic elements.

Contamination and the effects of hazardous waste on the environment can impact most ecological systems, especially ground water, soil, surface water and the air. Groundwater contamination is usually one of the greatest impacts and one that receives considerable attention.

The actual effects of contamination are varied as well. Disappearance of many control species and change in species' relationships occurs in some situations. Alterations in nutrient cycles, reduced functioning capability of some species and socioeconomic impacts may also occur [Wentz 1989].

Sampling and analysis can also be used to judge how well an organization is complying with environmental laws, regulations and internal policies. Site assessments or audits will help identify deficiencies and oversights that may result in a fine, penalty, or litigation.

This chapter is divided into three sections and presented in the order in which they would occur in real life. The first section, Sampling, provides information on sampling equipment and containers. The second section, Field Analysis, presents a few systems that allow one to get immediate results in the field. The final section of this chapter, Laboratory Analysis, involves the accurate analysis of samples under controlled conditions.

Sampling

Objectives

The primary objective of sampling is usually to collect a representative sample of the environment in question. The objective in some cases is to find out the highest concentration of the contaminant. Sampling also helps to determine compliance with regulations, isolate sources of contamination, determine the presence or absence of a compound, determine the amount of employee exposure, assess treatment or decontamination effectiveness, determine recycle suitability and estimate the adequacy of protection.

Sampling Plan

A sampling plan should be prepared before the sampling is started and cover the objective, what or who will be sampled, when to sample, the types of samples needed, the sampling locations and sample size. Entire books are written on determining the proper location and number of samples. The organic vapor analyzers discussed later help to determine sampling locations and can be a first phase of the plan. The sampling plan should also address the proper use of descriptions (such as soil descriptions), the number, frequency and duration of samples, decontamination procedures, waste management, collection and handling methods and the calibration of equipment.

The types of samples taken depend in large part on the medium being sampled. The three mediums commonly sampled for contaminants include air, liquids and solids. For example, in atmospheric/air sampling, area, source and personal samplers are used to assess environmental exposures. Personal/breathing zone samplers are used to measure occupational atmospheric exposures.

Liquid sampling would be of ground water, surface water, process streams and spills. Ground water is normally sampled via monitoring wells or by a hydropunch. Springs and underground mines have also been used to obtain ground water samples. Usually, surface waters sampled for hazardous waste are streams or rivers that receive discharges. Lakes and oceans are also sampled; however, it is much harder to obtain meaningful data from these larger surface water systems, especially in tracing contamination back to a particular source.

Ground water sampling in terms of hazardous contamination is a very complex topic. The number, location and depth of ground water sampling wells depends on many factors such as the soils, types of sources, geology and the contaminants.

When the sampling plan is being designed, the primary objective should be kept in mind, which is to define the contamination plume(s). In order to do that, it may be necessary to readjust the well locations and depths several times during design and operation, especially if the geology of the area includes fractured zones.

The challenge of getting meaningful data is still not over since the design, construction and operation of the wells will have a significant impact on the value of the data. Important considerations include cross-contamination of sample by sampling equipment, screen interval, slot size, casing diameter, intake depth, casing material, screens, well development and actual collection of the sample [Wentz 1989].

Solids sampling would be of soil, process materials and spills. Soils are commonly sampled after spills or leaks have occurred. The spill itself may be a solid and require sampling. Solids sampling equipment and protocol are vastly different from liquid and atmospheric sampling.

Hazardous vs. Nonhazardous Samples

There is an important distinction that should be made between hazardous sampling and nonhazardous sampling. A nonhazardous sample is reasonably known to contain no hazardous contaminants or low concentrations of such contaminants. These are commonly composite samples.

Hazardous samples, on the other hand, have unknown levels of contaminants or are considered to contain high concentrations of contaminants. They are usually collected from drums, tanks, lagoons or spills. Special safety considerations must be followed when sampling hazardous substances. In the majority of situations, the hazardous substance is taken as a grab sample. In the case of industrial hygiene, the grab sample is used to establish the peak and ceiling limits.

Figure 6-1 illustrates the proper use of personal protective equipment (PPE) to protect workers while collecting hazardous samples. Although not shown in the photo, PPE can also result in obtaining better sample results. An important factor in obtaining good results is that new gloves or other PPE are used for each discrete sample collected if they have become contaminated from the previous sample. The PPE being used by the individuals shown in Figure 6-1 includes coverall, full face respirator, gloves and boots.

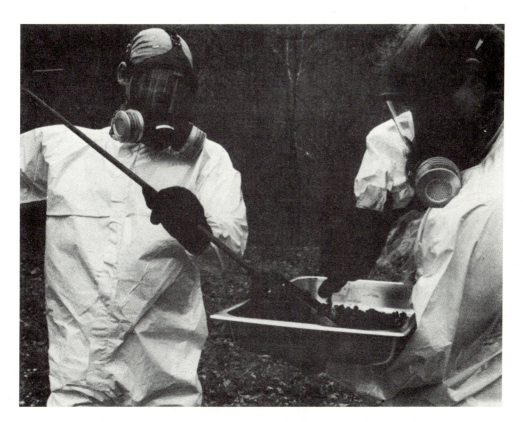

Figure 6.1 Collecting soil samples to measure possible dioxin contamination. Even though the EPA is reassessing its original dioxin cancer risk estimates, dioxin is still considered to be a highly toxic cancer-linked compound. A modern, affluent society offers many benefits, but also presents risks. (Reprinted from EPA, *EPA Journal* 13/9, Washington, 1987.)

Characteristics of Sampling Equipment for Hazardous Substances

Hazardous sampling equipment should have some unique characteristics. First, it should be easy to operate, since personnel may be wearing cumbersome safety clothing and respiratory equipment. It should be nonreactive so that it does not contaminate the samples. Closely related to ease of operation, it should be safe to use. Last, it should be inexpensive and disposable or easily decontaminated.

Examples of Hazardous Samplers, Containers, and Closures

Table 6-1 lists examples of hazardous sampling equipment. The equipment is arranged in four categories: liquids/slurries, sludges/sediments, powdered or granular solid, soil and air. This list only presents a few examples of the many types of sampling equipment available.

Table 6.1 Examples of Samplers for Hazardous Material/Waste

Category	Sampler
Liquids, slurries	Open tube (thief)
	COLIWASA
	Pond sampler
	Manual pump
	Powered pump
	Weighted bottle sampler
	Kemmerer sampler
	Extended bottle sampler
Sludges, sediments	Open tube
	Thin-wall corer
	Gravity corer
	Ponar dredge
Powdered or granular solids	Grain sampler
	Sampling trier
	Trowel/scoop/spoon
	Waste pile sampler
Soil	Soil auger
	Trowel/scoop/spoon
	Posthole digger/shovel/pickax
	Split spoon sampler

Table 6.1   Examples of Samplers for Hazardous Material/Waste (Continued)

Category	Sampler
Air	Passive dosimeter
	Midget impinger
	Fritted bubbler
	Glass bead column
	Spiral absorber
	Respirable dust sampler

Source: *USEPA Methods for Chemical Analysis of Water and Wastes.*

One item listed at the bottom of Table 6-1, respirable dust sampler, will be discussed as an example of a hazardous material sampler. This piece of equipment is composed of a small pump, hose and cassette placed as close to the operator's breathing zone as possible. The pump draws in a measured volume of air that would roughly equate to the amount the person would be breathing. The pump is left on for the entire shift of the worker. As the air is drawn through the cassette, large particles fall out into a cyclone, and the smaller or respirable particles are deposited onto a filter pad. The filter pad is later weighed to see if the operator has been exposed to a level of hazardous materials that exceeds the regulatory limits. Coal particulates are an example of a hazardous material that is sampled in this way.

A few examples of some of the hazardous waste sample collection equipment mentioned are illustrated in Figure 6-2. The suction system for sampling from waste containers, for example, is designed to obtain the sample without contaminating the worker or the pump. This is achieved by the use of two lines into the sample container. The suction line is near the top of the container, and the hazardous-waste dispensing line is near the bottom of the container. This prevents contamination of the pump unless the container is overfilled. A diastolic pump is used in a similar fashion, whereby the pump does not come in contact with the material or waste.

Once the sample has been collected, it must be put into the right kind of container. Table 6-2 lists the various types of glass and plastic containers and closures for hazardous samples. It is important that the proper container and closure be utilized, since some chemicals will dissolve plastic or glass. In other cases, the container will allow release of the contaminant, resulting in exposure or erroneous results. If glass is specified, it should be used since some chemicals will dissolve plastic. If, on the other hand, glass or plastic can be used, plastic is preferable because it will not break if the sample container is dropped. Some glass containers are plastic-coated to minimize

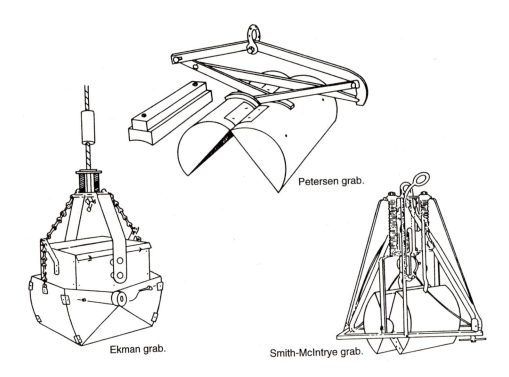

Petersen grab.

Ekman grab.

Smith-McIntrye grab.

Figure 6.2 Sampling Equipment (Part 1 of 3).

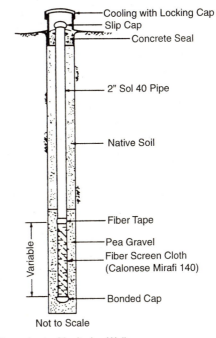

Cooling with Locking Cap
Slip Cap
Concrete Seal

2" Sol 40 Pipe

Native Soil

Fiber Tape

Pea Gravel
Fiber Screen Cloth
(Calonese Mirafi 140)

Bonded Cap

Variable

Not to Scale

Groundwater Monitoring Well

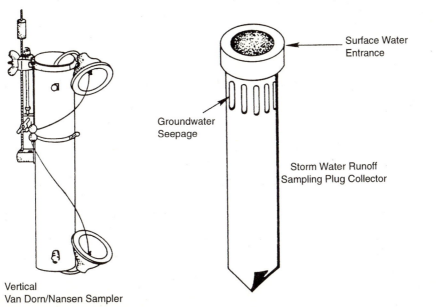

Surface Water
Entrance

Groundwater
Seepage

Storm Water Runoff
Sampling Plug Collector

Vertical
Van Dorn/Nansen Sampler

Figure 6.2 Sampling Equipment (Part 2 of 3).

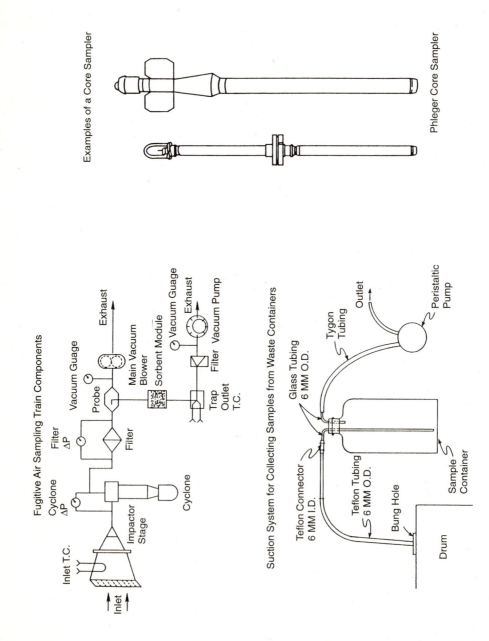

Figure 6.2 Sampling Equipment (Part 3 of 3). (From Dunlap, W.J. et al., *Sampling for Organic Chemicals and Microorganisms in the Subsurface*, Robert S. Kerr Environmental Research Laboratory, EPA-600/2-77-176.

breakage. Care should be exercised to ensure that the closure liner is in place and not damaged, which will happen frequently. If this occurs an exposure is possible or, at a minimum, the sample will not be accurate. Once a sample container has been selected, it is often necessary to preserve the sample to ensure accurate results. For instance, if a sample is to be analyzed for metals, a polyethylene bottle with a Teflon-lined cap would be used. The sample would then be preserved using a strong acid. The strong acid keeps the metals suspended in the sample to accurately reflect the waste in question. In addition, some samples require refrigeration.

Table 6.2   Sample Containers and Closures

Category	Recommended Container	Recommended Closure
Metals, inorganics, weak acids/bases	Glass or plastic	Plastic cap with plastic or Teflon liner
Organic solvents, hydrocarbons, chlorinated hydrocarbons	Glass	Plastic cap with Teflon liner
Photosensitive materials	Amber plastic or glass	Plastic cap with plastic or Teflon liner
Strong acids	Glass	Plastic cap with Teflon liner
Hydrofluoric acid, phosphoric acid	Plastic	Plastic cap with plastic liner

Source: *USEPA Methods for Chemical Analysis of Water and Wastes.*

Safety Plan

A safety plan is required any time hazardous samples are to be taken. Obviously, the safety plan helps to protect the individuals who obtain the samples and the general public. In addition, it has a side benefit of improving the quality of the data obtained. The plan must address the type of background, experience, training and certifications the workers must have. The PPE the samplers must wear should be clearly designated, along with any administrative controls that are needed, such as maximum number of hours doing the sampling. Concentration at which the PPE level must be upgraded must also be clearly spelled out. The samplers must have communication ability, such as portable phones, in case they run into problems. A place for decontamination, along with the proper equipment, must be spelled out. One key component of the safety plan, especially if old drums are involved, must be the proper opening procedure. For example, a spark-proof brass bung wrench should be used. If the drum has a bulging head, a remote opening device should be used.

Summary

Obtaining accurate, representative, legally defensible and meaningful samples is becoming a science on its own. There are many other aspects involved with sampling that have not been discussed, but are also important. These include topics such as sample area selection, number of samples, chain of custody, quality assurance and many other important subjects.

Once the sample has been taken, the sampler really has two choices: to either analyze the sample in the field using field analysis techniques or send the sample to the lab. The next section will address the former option, field analysis. The latter option, laboratory analysis, will be covered in the final section of this chapter.

Field Analysis

Introduction

Some samples can be analyzed in the field and an approximate level determined immediately. However, the rapid speed of obtaining a result in the field is partially offset by a drop in accuracy for most field instruments. This may be acceptable in certain situations, such as during initial site investigation, emergency response, sample location selection, sample screening and rough estimation of the extent of contamination. Field analysis also allows for monitoring of drilling discharge water to minimize the amount that must be placed in drums prior to disposal.

Organic Vapor Analyzers

Organic vapor analyzers are commonly used in the field to test for volatile organic compounds. Most of the detectors or analyzers are based on the ionization potential of the organic compound. There are two types of organic vapor analyzers, flame and photoionization.

Flame ionization detectors (FIDs) utilize a flame to obtain results. This causes a release of energy by breaking chemical bonds.

Photoionization detectors (PIDs) are commonly used to detect aromatic hydrocarbons. The PID cannot be used if methane and some other light organics are present, since these compounds cannot be ionized by ultraviolet light.

FIDs are useful in detecting organic vapors in air from 0–10,000 ppm, and PIDs detect 0–2000 ppm. Values are only correct for the specific gas to which they have been calibrated.

Magnetometry

A proton precession magnetometer is used to measure magnetic forces. Once the normal earth's magnetic field is accounted for, presence of buried ferrous metals can be inferred using this equipment. Ferrous metals create their own unique magnetic field which can be detected if the object is large enough, such as an underground storage tank or a cluster of drums. The accuracy of this method is dependent not only on the size of the metal object(s), but on the depth and soil characteristics as well.

Electrical Resistivity

Subsurface electrical resistivities in ground water, rock and soil are assessed in this system. An electrical field or current will move through materials, including contaminated areas at different intensities. Areas of low resistance can be shown, which indicate the possible presence of contamination plumes. The whole system depends on the application of electrical currents into the ground via surface electrodes.

Ground-Penetrating Radar

Subsurface materials with different electrical properties will sometimes reflect in their own unique way. In this system, high-frequency radio waves are introduced into the ground. A receiving antenna then picks up the reflected return wave. This equipment is expensive and normally used only for lighter-than-water contaminant plumes.

Test Kits

Test kits utilize indicator strips or reagents that are added to the samples in the field, allowing immediate results. Traditionally, the kits were developed primarily for water variables; however, this is changing. The kits usually contain test tubes, indicator strips, reagent and instructions.

The most common kits assess chlorine, pH, lead, chlorides, water solubility, cyanide, oxidizers, PCBs and chemical oxygen demand (COD).

Hazcat kits are a special kind of field test kit that have been developed and are used routinely by emergency response teams. These kits range in complexity from basic identification of a substance into four or five hazard classes to actually naming many chemicals. The basic kit indicates only whether the substance has characteristics of ignitibility, reactivity or corrosivity and whether PCBs are present. The more expensive Hazcat kits use a variety of procedures, such as colorimetric detection strips to actually identify certain compounds. Training in the use of such a kit is needed so that the field technician can quickly identify the substance. A few of the most important substances the Solids Hazcat is capable of determining include organophosphate, cyanide, phenol, arsenic and asbestos.

Portable X-Ray Systems

Portable x-ray systems are used when high levels of hazardous metals are expected in widespread situations. This type of system has been used, for example, by the mining industry. Portable x-ray fluorescence systems are used to identify the presence of certain metals such as arsenic, lead, mercury and others. This equipment is expensive and not accurate.

Dräger Tubes

Dräger tubes are used to measure contaminant levels in the air. By using a hand-held pump, a set volume of air can be drawn through various indicator tubes. These tubes contain substances that change color depending on the level of contaminant in the air. Examples of hazardous air contaminants that can be assessed using Dräger tubes include formaldehyde, ammonia, carbon monoxide, trichloroethane, hydrocarbons, methylene chloride, toulene, perchloroethylene, acetone and many others.

Portable Gas Chromatographs

Portable gas chromatographs are used to quantify or identify certain organic compounds. The technique has been used for many years in the laboratory setting with nonportable units. Over 33 compounds, such as benzene, can be assessed by drawing a gas sample into a probe. Retention time in the chromatograph column is measured against expected time. This field system is especially useful to screen samples that should be analyzed in the laboratory.

Other Field Meters and Systems

New field meters are introduced almost every day. These meters usually measure contaminants in the liquid or gaseous states and have varying levels of accuracy. Examples include meters that give immediate readings for pH, dissolved oxygen, lead, formaldehyde, ozone, temperature, oxygen and conductivity.

Laboratory Analysis of Hazardous Materials and Wastes

Introduction

Even if field analysis has been utilized in a specific situation, in most cases laboratory analysis is still done. Usually laboratory analysis is all that is done because of the higher level of accuracy. Also, if EPA methods are followed by a certified laboratory, the results can be used for waste classification.

Certified laboratories have strict guidelines for quality, tracking and control. For example, by using a chain-of-custody document, the laboratory can ensure that someone has responsibility and control of the sample from the time it is collected through analysis.

A certification process for laboratories has been set up by federal and state agencies to ensure that they meet minimum standards. To be certified, laboratories must show that they can perform the agencies' Best Demonstrated Available Technology (BDAT) for the specific contaminants they want to analyze. The laboratory must also present an acceptable quality control/assurance program.

The EPA has published a document that provides guidance to personnel involved with sampling and analysis. SW 846 details analytical procedures approved by the EPA for the identification and analysis of HMW. Many of these procedures are complex and require expensive equipment to complete analysis in a satisfactory manner.

Most of the HMW techniques used in the laboratory are based on the same analytical procedures used to analyze nonhazardous substances in the laboratory. The property of the chemical plays a large part in technique selection. The reputation of the technology is also an important consideration. This section will first present the laboratory techniques that are considered initial measurements. These will be followed by analysis techniques for inorganic contaminants and then organic contaminants.

Initial Laboratory Analyses

A wide variety of initial laboratory techniques for both inorganic and organic contaminants are utilized. Many of these miscellaneous techniques do not fit nicely into the categories mentioned below, but are valid techniques just the same. In many cases, but not all, they are done before the more detailed procedures. For example, laboratory meters and tests will provide gross (quick and less accurate) measurements of conductivity, pH, turbidity, total dissolved solids (TDS), dissolved oxygen (DO), chemical oxygen demand, oil and grease, biological oxygen demand, total organic halogens, total organic carbon and flash point.

Another group of laboratory techniques that could be placed into this initial category includes analyses for hexavalent chromium, nitrogen, phosphorus, sulfate, sulfite, sulfide, fluoride, chloride, cyanide and asbestos. Wet chemistry techniques are used where well-developed instruments and methods are lacking. Microscopic methods are also utilized, such as for asbestos, and can include phase contrast microscopy or transmission electron microscopy.

Detailed Analysis for Inorganic Contaminants

Atomic absorption spectrometry (AA) and inductively coupled plasma emission spectrometry (ICP) are two common laboratory techniques for analysis of metals. These two methods, or adaptations of these methods, are found in many laboratories. Additional detail on each of these two methods follows.

Atomic absorption The spectroscopic behavior of gaseous metal atoms is the basis of AA. The sample is atomized, and a light source is introduced. Characteristic wavelengths of light are absorbed by certain atoms. The light that is not absorbed is the component which is transmitted and the identifier of a particular atom. The quantity of the substance present is determined by the light's intensity. AA instruments have variations in specific aspects of their operation, but they all basically contain a detector, a wavelength isolation function, an optical mechanism, a computer and a light source. Specific types or adaptations of the AA include the flame, graphite furnace, cold vapor and hydride generation AAs. The flame and graphite furnace AAs are the most common. Each element is measured separately, hence making this a slower process than the next to be discussed.

Inductively coupled plasma emission spectrometry (ICP) In ICP, the spectroscopic behavior of gaseous metal atoms is also a basis of the procedure. However, in this case the sample is nebulized into a plasma with an input of energy from an argon torch. When the excited electrons relax and return to their normal energy state, they emit a characteristic wavelength of light. The two most common types of ICP include the simultaneous and sequential ICPs. The major difference between these two types is the number of photomultipliers used to transmit the light signal to the electronic converter. Many elements can be analyzed at one time, making this a much faster process than AA.

Detailed Analysis for Organic Contaminants

The gas chromatograph (GC) is the most widely used laboratory technique for analysis of organic contaminants. High-performance liquid chromatography (HPLC) is also used in many laboratories for organic analysis.

Gas chromatography The GC can separate a variety of organic compounds and identify them using several different types of detectors. GCs can assess organic compounds associated with halogenated and nonhalogenated hydrocarbons, aromatic compounds, phenols and chlorinated pesticides. In this method, samples are partitioned between a mobile gas phase, which percolates through a column, and a stationary solid phase in the column. The column selectively retards the sample components by adsorption or absorption. Either a packed or capillary column is used in GC. After the compounds are separated onto the GC column, they are removed from the column for identification and quantification by a detector.

There are numerous types of GC detectors. A few of them, including their specialties, follow: flame ionization (most organics), electron capture (chlorinated and ionized compounds), flame photometric (phosphorus and sulfur compounds), NP thermionic (nitrogen and phosphorus compounds), mass selective (most organics).

There are various limitations with GCs. For example, the samples must be cleaned up and volatile to be analyzed. The GC requires other instrumentation for confirmation of peak identity. Some classes of compounds cannot be analyzed, such as high-polar, nonvolatile, ionic compounds and high-molecular-weight polymers, without first being extracted with a solvent. This pulls the nonvolatile and high-molecular-weight polymers out of the sample so that they can be analyzed.

Figure 6.3 Times Beach. (Reprinted from EPA, *Environmental Data: Providing Answers or Raising Questions*, EPA Journal No. 15/3, Washington, 1989.)

High-performance liquid chromatography (HPLC)
HPLC is a relatively new methodology, but it is already starting to replace many GC methods. It involves a solvent reservoir, mixing system, high-pressure pump, sample inlet port, column and detector. The HPLC system has improved speed, resolution, sensitivity and column reuse ability. In addition, it can be used for separating the high-molecular-weight compounds, which have either a low vapor pressure or undergo pyrolysis when subjected to the higher required temperatures of GC [Gubber 1990].

Case Study

Times Beach, Missouri   The importance of analysis, preferably before the situation gets out of hand, is pointed out in this case study. Dioxin, one of the most toxic chemicals known to man, was involved in an incident in Times Beach, Missouri, which started around 1971. An estimated 48 lbs. of dioxin was dumped along roads and stables and other locations around Times Beach. On December 5, 1982, a tributary of the Mississippi, the Meramec River, flooded Times Beach and in the process spread the dioxin. It was known during the flood that the dioxin was being spread because soil samples had been taken one week before the flood. These samples showed dioxin levels as high as 100 ppb. The maximum safe level at the time was 1 ppb, according to the U.S. Center for Disease Control in Atlanta, Georgia. The EPA purchased all of the contaminated property [EPA 1989]. Figure 6-3 shows a load of contaminated furniture removed from homes in Times Beach. The furniture was hauled in covered trucks, which were escorted by the Highway Patrol, to disposal pits. The town was posted "Stay Out."

CHAPTER

SEVEN

Assessment

Introduction

The term "assessment" is being used in this chapter to refer to general investigative activities that are done to better understand the impacts of HMW. Sampling and analysis, which were just discussed, are some of the first steps in the assessment process. The property transfer assessment, operational audit and risk assessment all help to define the impacts of HMW on current operations. This chapter will concentrate on these three assessment types.

There are numerous types of environmental reports and assessments that traditionally have not concentrated on HMW. A few examples were discussed in Chapter 1. These included the environmental assessment (EA), the environmental impact report (EIR) and the environmental impact statement (EIS). Usually, these documents are required prior to construction of major projects (and regulatory changes) and deal with general environmental and socioeconomic impacts. Even if the EA, EIR or EIS are not discussed in this chapter, they are shown in Table 7-1 so that the reader can see the relationship between them and the assessments that will be discussed (i.e., the property transfer assessment—PTA, operational audit and risk assessment). Table 7-1 arranges the different assessment types from top to bottom in terms of timing (year 1 to year 20 or 30 of an operation) and from left to right in terms of severity of problems noted.

Table 7.1 Comparison of Several Common Types of Assessment

Timing	Name of Review	If Problem Noted
Prior to purchase	Phase I Property Transfer Assessment (PTA)[a]	Phase II & III PTA[a]
Prior to construction	Environmental Assessment[b]	EIR and/or EIS[b]
After operation for a period of time	Operational audit[a]	Risk assessment[a]
Prior to sale	Phase I PTA[a]	Phase II & III PTA[a]

[a] Emphasis of the assessment is hazardous materials or wastes.
[b] Emphasis of the assessment is project impacts.

Property Transfer Assessments

A property transfer assessment, due diligence review, prepurchase or pre-lease review are all similar terms. This type of review should be done for any commercial, industrial or multiple-family development prior to the purchase, lease, financing or sale of the land or building. This will minimize later impacts to the environment and reduce the risk of potential litigation, especially concerning hazardous waste or contamination left by a previous owner.

Environmental Liabilities in Property Transfer

Environmental liabilities during property transfer arise from various statutes. Of greatest importance is CERCLA, or Superfund, which establishes liability that is retroactive, eternal and can't be contracted away. In addition, SARA, RCRA, Clean Water Act, the Clean Air Act and other acts involve property transfer liabilities.

The level of liability that is involved in property transfer can be astronomical. The PTA will help minimize that liability. Some organizations that have not done a PTA have found themselves the owners of ground water cleanups costing many millions of dollars. Injuries and law suits can also be unwelcome aspects that accompany the deed to a property.

Due Diligence

Due diligence is a flexible use of processes and techniques intended to allow an interested party to evaluate the potential environmental risks associated with business transactions involving real estate. It is a risk management tool. It uncovers the likelihood of hazardous substance contamination on the property, the potential environmental liabilities associated with conducting the business or owning the property and the value of the property.

Due diligence developed primarily in response to CERCLA or Superfund. CERCLA states that parties who have had no involvement with the hazardous substance contamination of property may be found liable for very expensive cleanups merely because they own the property.

The extent of the due diligence survey is important. The survey must be consistent with good commercial or customary practice. It must be taken a step further when a reasonable person would be suspicious of environmental problems. The American Society of Testing and Materials has introduced standards for property assessments (E1527-93 and E1528-93). ASTM standards are often used in court, though there is no guarantee this standard will be adopted universally in terms of survey extent.

Innocent Landowner Defense

The innocent landowner defense is specified in the Superfund Amendment and Reauthorization Act (SARA) and is the only partial relief to the CERCLA liability problem. To qualify for the defense, the landowner must not have known that the property was contaminated at the time of acquisition. They must have made reasonable inquiries into the past uses of the property before acquisition to determine whether the property was contaminated (environmental due diligence). The property must have been acquired by the defendant after the disposal or placement of the hazardous substance. They must have reacted responsibly when the contamination was found. As a practical matter, qualifying for the defense is very unlikely under most circumstances.

In addition to the innocent landowner defense, there is an exemption for parties that hold title to property because they have a security interest in the property. The EPA has published a regulation indicating what actions these parties can take without losing their protected status (40 CFR 300).

Levels of a Property Transfer Assessment

There are varying degrees of depth at which a property transfer assessment can be done. These range from a very general overview to an in-depth study. The different levels are shown in Table 7-2 and include an initial screening and Phases I, II and III. In general, an initial screening does very little; a Phase I identifies potential problems; a Phase II defines the problems; and a Phase III provides costs and correction. Phases I, II and

III are defined very loosely, so it is important that the consultant spells out very clearly what will be covered in each phase.

Table 7.2 Levels of Assessment

> ■ *Initial Screening*—Quick visual overview
>
> ■ *Phase I (Tier I)*—Identification of potential problems
>
> ■ *Phase II*—Characterization of identified problems
>
> ■ *Phase III*—Detailed cost analysis (and corrections)

Public Records Review

Table 7-3 presents some of the records that should be reviewed during the record review portion of a Phase I property transfer assessment. Of all the records mentioned, the federal (see II.A.1) and state (see III.F.1) Superfund lists and regional agency lists of leaking tanks (see IV.B) are the most important. Aerial photographs and Sanborne Fire Insurance Maps are the most informative for historical information. Old city directories that are arranged by street address are also very useful, but have only been prepared for certain locations and years.

Table 7.3 Public Records Review: Due Diligence Checklist

> I. Real Estate Records
>   A. Title Reviews
>   1. Grantor/grantee 50 year chains (or longer if records are available)
>
> II. Federal Sources of Information
>   A. EPA (and in some cases state)
>   1. Superfund National Priorities List—NFL Sites
>   2. Comp. Emergency Response, Compensation and Liability Information System—NFL sites and sites which may be contaminated
>   3. CERCLA Release Notifications and Follow-up Reports
>   4. Emergency Remedial Response Information Center System—sites potentially contaminated
>   5. Toxic substances Release Inventory—releases of a toxic chemical
>   6. TSCA Inspection and Compliance Records—PCB violations
>   7. Violator List—Violations of various environmental laws
>   8. NPDES and POTW approved discharges
>   9. Petroleum and Oil Spill Reports
>   10. UST Records
>   11. EPA ID number—generators
>   12. Asbestos notifications about demolition, renovations, and releases

Table 7.3 Public Records Review: Due Diligence Checklist (Continued)

       B. U.S. Army Corps of Engineers
           1. Section 10 permits
           2. Section 404 permits
       C. USGS
           1. Ground water information
       D. Security and Exchange Commission
           1. Filings regarding potential liabilities
       E. Emergency Planning and Community Right to Know Agencies (Fire Department, EPA, State)
           1. Stored hazardous chemicals and extremely hazardous waste

III. State Sources of Information
       A. Air Quality
           1. PSD permits
           2. Air construction permits
       B. Water Quality
           1. NPDES permits
           2. Enforcement actions
       C. Coastal Zone Management
           1. Building permits
           2. Wetland issues
       D. Natural Resources
           1. Oil, gas, and mineral development permits
       E. Solid Waste Management
           1. Disposal of nonhazardous solid waste
       F. Hazardous Waste Management
           1. TSDF permits and Superfund list
           2. Audits and Enforcement violations
       G. Fish and Wildlife
           1. Damage records

IV. Local Sources of Information
       A. County, City, Fire Departments
           1. Building permits
           2. Land use permits
           3. UST —HMMP
       B. Regional Water Quality Control Boards

V. Other Sources of Information
       A. Aerial Photographs
       B. Computer Data Firms

           Aerial photographs were one example of a source of the historical information just mentioned. Figure 7-1 is an aerial photograph taken on July 23, 1980. By studying the original photo, especially with magnification and in stereo, the industrial operations, waste disposal areas, above-ground storage tanks and other surface features can be noted. Sometimes, it is even possible to distinguish stained soil, stressed vegetation and unauthorized drum-dumping areas.

Figure 7.1 Aerial photograph. (Pacific Aerial Surveys, 1980. 8407 Edgewater Dr., Oakland, California 94621)

Field/Site Review

Table 7-4 lists some of the things to look for during the field or site review portion of a Phase I [Denton 1989]. The field inspector must be trained to look for subtle clues of present and past contamination. Most HMW real estate problems of importance are not going to be obvious.One item not shown in Table 7-4 that would be important to check is the drawings for the operation going all the way back to the first-year grading/earthwork drawings. This may show that, a few years ago, a sump, tank, trench or other structure was installed. It may now be out of service and covered. It also may have leaked when it was being used. Usually, there would be no way of seeing either the structure or leak by walking the site. This makes a review of the past drawings very important.

Table 7.4   Site/Field Review: Due Diligence Check List

I. UST and Piping—Check HMMP and building permits for:
    A. Age
    B. Type—Problems with bare steel tank and piping
    C. Spills and Overflows

II. Asbestos—Check for:
    A. Demolitions and renovations
    B. Releases
    C. Presence of ACM
    D. Age of building

III. PCBs
    A. Age of electrical equipment
    B. Electrical work areas
    C. Electrical disposal areas

IV. CERCLA Hazardous Substances
    A. Any amount of the 717 substances listed in 40CFR Table 302.4

V. Processes
    A. Industrial
        1. Type of past and present processes
        2. Permits adequate, current, acceptable limitations and constraints
        3. Pollution control equipment o.k.
        4. Grandfathering of equipment
    B. Waste Management and Disposal—Compliance of past and present processes and waste streams
        1. Treatment and disposal into POTW
        2. Treatment and disposal via NPDES
        3. Disposal via Class I, II, or III
        4. Recycling or Reuse
        5. Has waste minimization been practiced?
        6. Are environmental systems at capacity or is there some growth potential?
        7. What kind of management support exists?
    C. Obligations for Mitigation Efforts

Table 7.4 Site/Field Review: Due Diligence Check List (Continued)

VI. Wetlands and Flood Plains
    A. Proximity to Water
    B. Frequently Inundated with Water
    C. Army Corps of Engineers Permits
    D. Fill Activities in the Past
    E. Dredge Activities in the Past

VII. Hazard Communication Program Records
    A. List of hazardous chemicals
    B. MSDS
    C. Proper handling, warning, emergency planning

VIII. Radon
    A. Short term test or "grab sample" (immediate)

Types of Facilities Requiring Review

A Phase I should be done prior to any real estate transaction involving all types of commercial, agricultural, industrial and multiple-family residential properties. The depth depends on the type of operation. At the present time, a Phase I is not strictly required, just inferred or recommended. However, most real estate transactions require the seller to fill out a property disclosure checklist, listing knowledge of environmental problems.

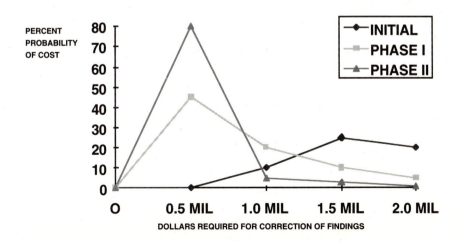

Figure 7.2 Property Transfer Assessment: Cost to Correct Findings.

Costs Identified During a Property Transfer Assessment

Figure 7-2 summarizes all phases of a property transfer assessment in terms of cleanup cost for a site in Illinois. As can be seen, the initial screening does not provide much information, especially in terms of cleanup cost, and is a more conservative estimate. The Phase I assessment starts to show some clarity in cleanup cost (range from $.5 million to $1million). In this case, the Phase II assessment nailed down the cleanup cost at about $500,000. Since the curves never meet the *x*-axis, it must be understood that there is always a chance that the costs could go lower or higher.

Operational Audit

In the preceding section, we were at a pre-lease, preloan, prepurchase point in time. In this section, the organization is further along in time and has already purchased or leased and has operated at the site for a period of time. It is now time to analyze or audit the operation in terms of ongoing impacts and compliance. This includes the organization's own operations and contractor operations.

From a regulatory standpoint, it is as critical to do a routine compliance check on operations as it is for an individual to have a routine physical exam. With laws and regulations changing almost daily, an operation can quickly move into noncompliance. In addition, the U.S. EPA has a formal policy encouraging the development, implementation and upgrade of auditing programs. The EPA will take into account the existence of such programs when determining the appropriate penalty in the event of a violation. In certain EC countries in Europe, the eco-audit regulations require that operations perform audits as of January 1, 1994.

The suggestions presented in this section could apply to contractors as well as an organization's own operations. There should be a good agreement in place with the contractor that protects the organization. The agreement should hold the contractor responsible for compliance with all environmental, health and safety laws and regulations. In addition, the contractor should have considerable environmental insurance or adequate self-insurance capability. The agreement does not completely remove the necessity to do occasional contractor audits. The audits should not violate the normal "independent contractor" condition in most agreements.

An operational site analysis or audit is a systematic assessment of compliance. It is the process of determining whether all or selected levels of the organization are in compliance with the regulations, internal policies and accepted practices. It is a check of the regulatory status of an individual facility.

Levels or Degrees of Depth

Different organizations invest varying degrees of time in operational audits. This is dependent on issues such as past fines, available resources, time and number of hazardous waste streams. The following four levels of audit are recommended, in decreasing order of importance:

- Known contaminant impact analysis—If a problem is known or suspected an in-depth analysis should occur to determine the quantity and extent type of contamination.

- Regulatory compliance analysis—A regulatory compliance analysis should be done by all organizations. This involves an assessment of compliance with all HMW laws and regulations in reference to a specific site.

- Corporate policy compliance—An assessment of corporate compliance should be done by most organizations, if possible, assuming that they have established HMW policies over and above regulatory requirements.

- Best management practice compliance—Best management practice for the industry should be compared to an organization's practices if significant hazardous waste streams are present. This is done in order to stay ahead of the regulations. It reflects generally what the rest of the industry is doing about specific hazardous wastes, especially in terms of treatment. Best Available Control Technology (BACT) is one way of describing treatment levels.

Key Principles

Many principles should be considered when designing an operational audit system. For example, what is the objective, and is it achievable? Will the system be systematic and supported by evidence? Results must be reported understandably, quality assurance/quality control (QA/QC) followed, and the field individuals must be involved in the process. The analysis must be fair, measurable and done in a nonthreatening way. It should not contain opinions or judgments. The analysis should be done at a set frequency, with some surprise visits as well.

It is important to consider the legal ramifications of audits. For example, if problems are uncovered and documented, they must be promptly corrected. If they are not, these "smoking guns" can do an organization great damage. In other words, a documented problem that is not corrected is a prime candidate for a fine or court action.

Total environmental quality management (TEQM) audits are an attempt to bring total quality management principles to bear on environmental problems. TEQM was pioneered by the Global Environmental Management Institute. In TEQM audits, the primary focus is on systems rather than specific compliance. For example, in the traditional audit, the inspectors review all HW drums for proper labelling and then document any deficiencies. In the TEQM audit, they take the additional step of asking whether systems are in place to ensure that HW drums are labelled 100% of the time. With this emphasis, the people being audited are less likely to look at the audit team as nitpickers from the home office and are, in fact, an important part of the audit team.

Audit Team Composition

There should be careful thought concerning who should audit, since different situations dictate different types of assessors or auditors. First, the auditors should be qualified and have a good understanding of internal processes and controls. If one does not understand the internal controls, it is extremely hard to verify that they are functioning. It may be appropriate for the auditors to be from an agency or a consultant, company staff, or field organization. They could be either part-time or full-time. However, it is hard for the part-time auditor to stay on top of all the applicable laws and regulations. Each situation usually dictates a different type of auditor or audit team. In the author's experience, an internal audit team should be composed of at least one regulatory expert and one field representative who knows the site and the process.

In some parts of the world (Europe, for example), there is proposed legislation that would require that an independent auditor assess for operational compliance with hazardous waste regulations. If the assessment is done, corrections made and results offered to the public, then the operation can obtain a "green" status or bill of health.

On an even broader scale, the ISO quality certification is starting to include environmental policies and procedures. Similar to the above example, an independent auditor must assess a company operation; if it meets quality criteria, it would obtain a certification. This certification and the above "green" status are very important for staying in business in Europe and are starting to become an issue in the U.S. as well.

Common Analysis/Audit Tools

There are some tools that are used during most audits. Usually, an auditor will select one or two that best fit the situation. Some of the choices include a topical outline (lists topics to be covered), detailed guide (lists regulatory requirements plus standards), yes/no questionnaire, open-ended questionnaire (explanation required) or scored questionnaire (responses are scored against criteria). No matter which tool is selected, it is important that the auditors use the tool. Far too often, auditors will "wing it" or just walk the site and expect their memories or experience to remind them what to look for. This is a risky habit to get into, since there are hundreds of things to consider.

Table 7-5 lists some possible HMW issues that would be assessed during a typical audit. In addition to the suggestions listed, state and local requirements should also be included. This list is not meant to be all-inclusive, just to provide some examples. Actual audit lists that the author has used in the past range from two pages for an administrative office up to 20 pages for a large manufacturing facility. Most audit lists would also include columns for suggested corrective action, scheduled completion date, actual completion date and responsible individual.

Table 7.5 Operational Audit: Examples of Hazardous Material and Waste Issues to Assess

Air Pacs	Air pacs (SCBA) are full, inspected monthly and tagged
Alarms	Evacuation alarms are present
Audits	Hazardous waste treatment workers audit & record daily
Audits	Supervisors complete and document general safety and hazardous waste storage/treatment area weekly
Chemical Labels	All chemical containers/vessels are labeled with hazardous contents and appropriate warnings
Chemical Exposures	Medical surveillance and IH data is collected for employees exposed to anything that may approach PEL
Chemical Handling	Chemical handling is centralized
Chemical Inventories	SARA Title III inventories are completed and submitted on schedule each year

Table 7.5   Operational Audit: Examples of Hazardous Material
and Waste Issues to Assess (Continued)

Chemical Spill Audits	Weekly chemical audits are completed and documented
Chemical Storage Secondary Containment	Chemical storage areas have secondary containment or absorbents
Chemical Storage Segregation	Incompatible chemicals aren't stored together
Chemical Storage Room	Room or area is well ventilated and clearly posted with signs
Chemical Transporting	Chemicals in glass containers are transported on-site via carts or rubber carriers
Chemical Volumes	Quantities stored comply with UFC. No more than 1 day supply of flammables or corrosives outside of the main storage area
Complaints	A procedure is in place to handle HMW complaints
Compressed Gas	All cylinders are capped and secured
Concealed Hazards	Significant concealed hazards are disclosed to affected employees immediately. Faulty system removed from service, repaired or OSHA notified in 15 days
Containers	No open containers of chemicals or waste
Discharge	No discharge to streams, drains or sewers unless analysis on file that shows compliance
Disciplinary Action	Applied for violation of laws and procedures
Dust Collection	Respirable dusts are vented to a collection system
Emergency Action Plan	A written plan is in place to address spills, tank ruptures, etc.
Emergency Coordinator	An on-site person has been designated
Emergency Notification Numbers	Numbers are posted throughout and in all required documents (manifests, HMBP, etc.)
Emissions Reporting	All required discharge data is reported
Empty Chemical Containers	Residual chemicals are used and container is washed or returned to supplier
EPA ID Number	On file for the site and used during manifesting
ERT Drills	ERT performs monthly emergency drills
ERT Plans	Plan contains current names, numbers, etc.
Evacuation	Adequate, up-to-date evacuation maps are posted
Eye Protection	Eye protection is available where chemicals and wastes are handled
Eyewash/Shower	Functioning unit is within 25' of chemical handling or storage areas

Table 7.5 Operational Audit: Examples of Hazardous Material and Waste Issues to Assess (Continued)

Fire Extinguishers	Extinguishers are within 10 feet of flammable storage
First Aid Assistance	A nurse or EMT is available on-site
Hazardous Waste Storage Time	Waste is not stored beyond the 90 day, 180 day or 365 day period specified
Hazardous Contingency Plan	A management or business plan is on file which deals with chemicals, wastes and emergencies
Hazardous Waste Storage Communication	Internal and external communication is provided in storage and treatment areas
Manifests/Records	All hazardous waste has a properly prepared manifest and LDR forms. Exception reports, biennial reports and waste determinations are kept on file for at least three years
Material Safety Data Sheets	MSDSs are available at the location where chemicals are present and employees are aware
Notifications	Employees are notified of chemical exposure levels. Medical files are up-to-date
Permits/TSDF	A TSDF permit or permit-by-rule is on file if hazardous waste is stored over 90 days or treated on-site
Permits/Water Discharge	All necessary permits for discharge to streams, storm run-off and sewers have been obtained
Personal Protective Equipment	Surveys, fit testing, training and follow-up done to insure proper use of PPE
Pumps and Pipelines	No leaks are present and they are checked on a routine basis
Respiratory Protection Program	Appropriate respirators are present and maintained after surveys, fit testing and training
Risk Assessment	An assessment of risk has been performed where there is potential significant risk
Signs	Required signs are posted, such as Hazardous Waste, etc.
Spill Clean-up Equipment	Absorbents, neutralizers, gloves, etc. are present where fuel, chemical and waste are used or stored
Sumps	Sumps and secondary containment berms are sealed, lined, free of liquids and inspected routinely for cracks and so forth
Training	Employees are properly trained and tested in chemical handling, hazard communication, first aid, etc.
Vendors	Hazardous waste vendors are registered and checked-out prior to use and at routine intervals thereafter

Table 7.5 Operational Audit: Examples of Hazardous Material and Waste Issues to Assess (Continued)

Ventilation	Adequate exhaust ventilation exists where needed
Waste Characterization	Wastes have been tested and there is an analysis on file regarding whether waste is hazardous
Waste Minimization	A pollution prevention program is in place

Audit Process Steps

The actual audit process steps range from very simple and straight-forward for small sites without many waste streams to very involved procedures for large facilities. The following generic checklist of activities would apply, in general, to both ends of the spectrum. The amount of time spent in each phase could range from several hours to several days.

Pre-audit activities   There are some actions that should be done before visiting the site, by the audit team. These include selecting the audit team, selecting facilities (mini-risk assessment), scheduling the audit and gather background information for the team (paper search). Most important, it is necessary to contact the site management before the audit. Their schedule, sensitivities and needs must be considered for the audit to be successful. It may even be advisable to not refer to it as an audit, since this word holds a negative connotation for many people. It could be termed an "operational review" or "compliance check."

Field analysis/audit   Once the upfront activities are done, the field work can start. The following actions are done at the site by the audit team and site individuals. The first action is usually a kick-off meeting at the site. This lets everyone know the purpose and schedule for the day. Next, interviews and other on-site information gathering occurs. This is followed by the most important step, the location walk-through. Audit findings are entered into a checklist during the walk-through. The final step at the site is an exit meeting to review and summarize the initial findings.

Post-audit, followup and correction   Post-audit actions are done after the field work by the audit team and site individuals, with as much involvement of the latter as possible. The audit team will first analyze and interpret the findings. The checklist or action plan is usually prepared to commit to specific actions by specified times. If they are big problems, a feasibility study and remedial action plan may be necessary. The audit team should then distribute the checklist or action plan. Far too often, it all stops here. Uncorrected findings (smoking guns) can lead to real liability problems. Correction and followup to ensure that the problems and their root causes are fixed are of paramount importance. An example of a followup checklist form can be found in Table 7-6.

Table 7.6   Audit Action Item Followup Checklist

Problem	Suggested Correction	Proposed Completion Date	Actual Completion Date	Corrected by Whom

It is important that even if a long report or action plan is prepared, a checklist is also completed. OSHA has promoted the use of checklists. It is much easier for the site individuals to fix a list of items, each with their own correction action, date and person assigned. It is also easier for the auditor to track this type of checklist to ensure completion. Auditors should not prepare long paragraphs full of various actions, all of which are to be completed by unknown individuals on one unrealistic date. It is an open invitation for failure.

Examples of Common Problems Noted in Audits

The types of problems uncovered in audits are numerous and could fill several volumes. The author has noted the following trends after working in a variety of industries: lack of training or understanding, lack of control and tracking systems, inadequate facilities, lack of management systems, higher than acceptable employee or public exposure, procedures inadequate or not understood, environmental pollution occurring, hazardous material source reduction not practiced, improper storage of hazardous materials and missing permits, plans and variances. Record-keeping deficiencies are a common problem, along with lack of resources (money and staff).

Ongoing Monitoring

An operational audit, a permit requirement or various regulations may require an organization to do ongoing monitoring. This may involve ground water, process discharge, worker exposure, surface water and many other types of monitoring. Only by monitoring can an organization really know where they stand in terms of protecting the environment and workers' health. This also allows the organization to make corrections early, before problems get out of hand. Please refer to Chapter 6, "Sampling and Analysis," for details that apply to monitoring.

The only self-monitoring that will be discussed further is for ground-water monitoring. All TSDF permit holders are required to have sufficient wells to assess the quality of background water and detect any possible contamination from the operation. Indicator parameters, waste constituents and reaction products must all be measured.

Risk Assessment

Many organizations are starting to do risk assessments in reference to some of their HMW. Certain regulations, such as California Prop. 65, require a warning if there is significant risk of exposure. The risk assessment allows this exposure determination to be made. Indirectly, risk assessments are also required by CERCLA, SARA, the National Contingency Plan and other regulations.

The discussions in this section concentrate on health risk assessments. There is, however, a whole new field developing: ecological risk assessments. As the names imply, health risk assessments estimate the impact of a chemical(s) on a human population, while ecological risk assessments estimate the impact on the entire ecosystem. The two types are compared in Table 7.7.

Table 7.7 Risk Assessment Comparison

Health Risk Assessment	Ecological Risk Assessment
■ Hazard Identification	■ Receptor Characterization (species, life stages, etc.)
■ Dose-Response	■ Hazard Assessment (nature and toxicity of effect)
■ Exposure Assessment	■ Exposure Assessment
■ Risk Characterization	■ Risk Characterization

Risk assessment helps to estimate the likelihood of injury, disease or death resulting from exposure to a potential hazard. Since there are many unknowns and uncertainties in a risk assessment, there is a need for more uniformity and standardization. The four main elements of a health risk assessment are illustrated in Figure 7-3.

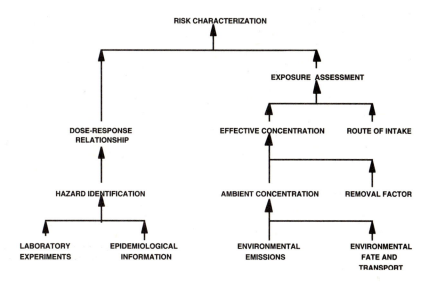

Figure 7.3 Risk Assessment.

Hazard Identification

In this stage of risk assessment, one would review, analyze and weigh toxicity and carcinogenic effect data in various settings. Just as the title implies, the goal of this first step is to identify the hazard(s) or base cause of the problem. Some researchers refer to this first step as the toxicological evaluation. It is a qualitative evaluation to determine whether the chemical may have a potential adverse effect.

Types of studies Human, animal and cell culture studies are the sources of the raw data. Human studies utilize case reports and epidemiological studies. The epidemiological studies are better but still not controlled, since they are usually from an accidental exposure.

In animal studies, mice, rats, rabbits and other animals are used and then one must extrapolate to humans. Sometimes the correlation is low between species. Animal studies include acute studies (short-term, high-level exposure) and chronic dose studies (life-time, low-level exposure).

Cell culture studies utilize test tubes or flasks containing cells. The cells are usually attached as monolayers or free-floating. For example, it is possible to get monolayers of skin and liver cells to grow in cultures. The HMW is then introduced onto a specific number of growing cells.Extrapolation has to be to the whole organism, which is hard to do since there is no organ interaction. One advantage of cell culture studies, however, is that impacts on cellular functions can be seen fairly quickly. Another advantage is that it reduces the amount of animal studies necessary along with the associated animal suffering.

Types of effects to analyze   During the above-mentioned studies, many observations are made. The most important categories of effects are usually mutagenesis, carcinogenesis, teratogenesis, weight loss and liver damage. Irritation, asphyxiation and sensitization are also effects that are commonly assessed.

The carcinogenic effect is usually the driving factor in risk assessment, because many people feel there is no threshold dose (as little as one molecule may cause cancer). In addition, a high degree of uncertainty requires low acceptable or worst-case limits. One way to classify carcinogens is by the weight of evidence system, which includes known, probable, possible, questionable and negative evidence.

The Dose-Response Relationship or Evaluation

The next phase of risk assessment is the dose-response evaluation. At what dose is the substance toxic or carcinogenic? The purpose of this phase is to estimate the incidence of adverse effect as related to the magnitude of human exposure to a substance.

The most important part of a dose-response evaluation or curve is the lower dose area, an area where few if any data points exist. To address this problem, many mathematical models have been developed. The mathematical models are used for extrapolating data to the important lower doses. All of the carcinogen curve models illustrated in Figure 7-4 are non-EPA-approved except the Linear Multistage Model [Wentz 1989].

The Linear Multistage Model is approved by EPA because it is the most conservative. This model assumes that there is no threshold or reference dose and that any amount (even one molecule) will cause cancer.

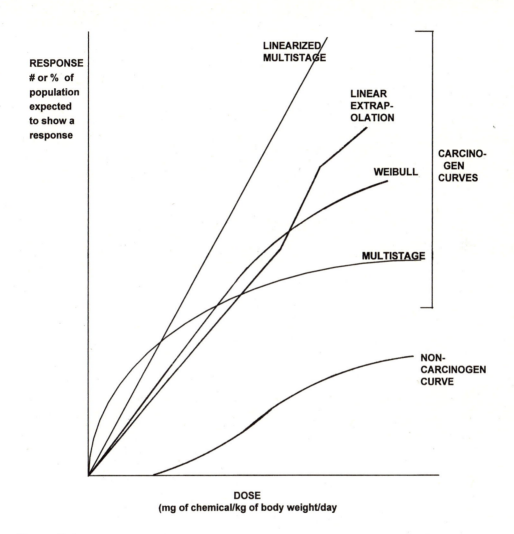

RESPONSE
or % of
population
expected
to show a
response

LINEARIZED
MULTISTAGE

LINEAR
EXTRAP-
OLATION

WEIBULL

MULTISTAGE

CARCINO-
GEN
CURVES

NON-
CARCINOGEN
CURVE

DOSE
(mg of chemical/kg of body weight/day

Figure 7.4 Dose-Response Curves.

The noncarcinogen curve is different from carcinogen curves in both shape and where it crosses the *x*-axis. A reproductive toxicant, for example, would have a dose below which there would be no measurable response. This is called the no observable adverse effect level (NOAEL).

Exposure Assessment

In this phase, one would ask the question how much human exposure may be expected? Agencies such as the EPA are especially open to negotiation at this phase of risk assessment, since there is considerable room for interpretation concerning the chance that humans will be exposed.

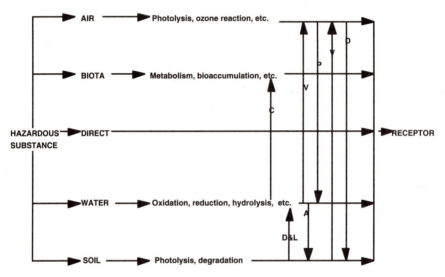

V = Volatization, C = Consumption, A = Adsorption, P= Precipitation, D = Desorption, L = Leaching/runoff

Figure 7.5 Exposure Assessment/Transport and Fate.

First, the source(s) and potential contaminant migration pathways should be analyzed. This helps determine where and how the chemical goes after its release from the source or the environmental fate while it is traveling to the target receptor. The target receptor may be a human, bird, plant and so on. The chemical may be transported in air as a vapor, dust or gas. It may also be transported in water (surface, groundwater, run-off). Transport in plants and animals also occurs, with bioaccumulation in some cases. Last, transport in soil is also a migration pathway. These pathways are summarized in Figure 7-5. As the figure illustrates, a contaminant may move between the major pathways. For example, the process of volatization may take a volatile organic compound from the soil pathway to the air route.

Figure 7-6 "Exposure Pathway Related to Ground Water," is a more in-depth illustration of just part of one line, the water pathway of the overall transport and fate illustration presented earlier. Depending on the contaminant, some of the other components should be broken down as was done for ground water. With this amount of depth and detail, computer programs have been designed to model a contaminant through all the exposure pathways and subpathways.

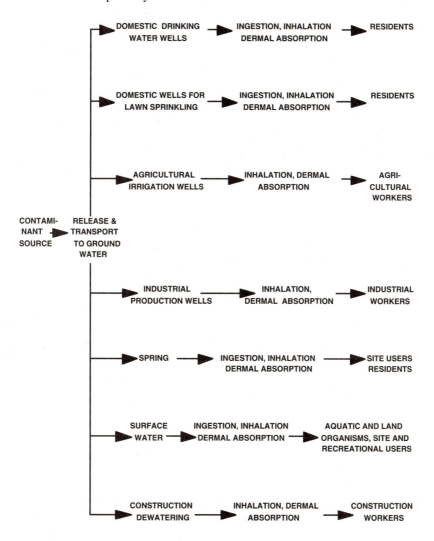

Figure 7.6 Exposure Pathways Related to Ground Water. (From EPA, *Guidance on Remedial Actions for Contaminated Ground Water,* EPA/540/G88/003, Washington, 1988.)

Figure 7-6 has even missed a few subpathways. For example, during well drilling it is possible that a well will go through two or more independent aquifers. If a well is drilled, constructed or abandoned improperly, it can create a subpathway where contamination is spread between previously independent aquifers. Although it does occur, cross-contamination is limited because of current well construction standards and the awareness of well drillers and geologists of the potential implications. When it does happen, however, transfer of contaminants in this exposure pathway can occur during all types of well drilling, including those shown in Figure 7-6 for domestic, agricultural, industrial, construction and one type not shown, ground water sample wells!

It is important to identify the human exposure points. These include inhalation in air, shower mist and cooking mist. Dermal contact (including injection) is the second type of exposure point. Ingestion in food, water and soil is the last type of exposure point.

An estimate of exposure level or exposure point concentration is the last part of this phase. This can be done by either monitoring and/or modeling. The estimate must consider the extent and frequency of exposure, or how much and how often (short-term or chronic exposure). The amount of absorption, excretion, metabolism and other physiochemical factors should also be factored in. Other variables to consider include the number of people exposed, variation of exposure, degree of intake by various routes, health and age of humans considered, background levels and interaction with other pollutants.

Risk Characterization

The last phase of risk assessment asks, what is the individual or the cumulative population risk? What is the likelihood of injury, disease or death resulting from human exposure to a potential hazard under the conditions assessed during the first three steps of the risk assessment? This is usually expressed as noncarcinogenic or carcinogenic risk. Risk characterization can be calculated for one exposure or cumulative exposures for either an individual or an entire exposed population.

There are several good references on risk assessments. The following are all EPA documents: Risk Assessment Guidance for Superfund Sites, Review of Ecological Risk Assessment Methods, Ecological Assessment of Hazardous Waste Sites and Guidance for Ecological Risk Assessment.

CHAPTER

EIGHT

Spills, Leaks and Site Contamination

Introduction

Contamination of water, soil and air can occur in an infinite number of ways. In the first section of this chapter, emergency response to the tipped-over drum or the highway/ railroad type of spill will be addressed. The second section deals with oil spills, and the third section of this chapter will present information concerning the non-emergency handling of leaks and site contamination (incidental releases). The final section of Chapter 8 will discuss CERCLA and SARA. This chapter deals primarily with U.S. regulations that apply to spills, such as CERCLA. A brief mention will be made, however, of the general situation in Europe and the Far East.

Figure 8-1 shows where most of these problems occur. 30% of the cleanups are associated with production facilities, 17% at waste management or TSDFs, 16% are midnight dumps from unknown sources and only 1% are transportation-related. The "other" category, which accounts for 36%, is made up of incidents such as spills at small quantity generators, small businesses, maintenance shops, construction sites and so on.

Presently, there is no comparable CERCLA/Superfund in Europe or the Far East. It is commonplace for many operations to sit on top of soil and ground water contamination caused by a past bankrupt operator. The current operator in most of those cases is not too concerned because the regulations do not hold them responsible unless they are presently adding to the contamination.

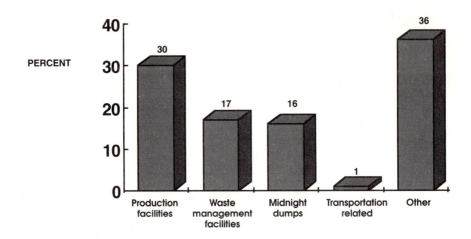

Figure 8.1 Removal Actions by Type of Incident. (From *EPA, CERCA/Superfund Orientation Manual*, Washington.)

If the past contamination is so bad that it impacts drinking water, the government cleans it up, not the current owner/operator.

This situation may change soon for Europe. On March 17, 1993, the EC Green Paper of Remedying Environmental Damage was published. The concept behind the paper is to cause more polluters to pay through civil and strict liability. It is interesting to note that a Green Paper objective is to not develop a litigation industry similar to that in the U.S. concerning Superfund.

Emergency Response to Spills and Hazardous Incidents

Everyone associated with HMW, especially emergency response (ER) teams, HAZMAT teams and transporters, should know how to act and where to quickly obtain assistance when hazardous material accidents occur. The key to success is knowing who to call for help and what type of assistance is available. The Federal Emergency Management Agency in Washington is a good source of this type of information.

An incident command system is used in many cases to allow responders and agencies to work together toward a common goal. It uses common terminology to allow ERs to communicate efficiently.

Emergency response to spills is an involved and complicated situation. Each occurrence is different; therefore, the following generic steps are only suggestions and can never be strictly categorized or ranked.

Emergency Response Steps Any Knowledgeable HMW Person Should Take

1. If the spill is at a facility (vs. a highway), call the site ER team.

2. Put up warning signs or barriers.

3. Keep unauthorized people back. Detour traffic and isolate area.

4. Observe at a safe distance. Look for things such as labels, placards, container type, ID numbers, safety relief valves and insulation. Question the driver.

5. Notify the police, fire and National Response Center at (800) 424-8802. The transporter is required to notify the NRC if the spill exceeds the reportable quantity (RQ).

6. Consult any available documents such as DOT's Hazardous Materials—Emergency Response Guidebook, MSDS, Business Plan and ER Plan.

7. Only if it is absolutely safe to do so, remove the injured individual and stop or slow the leak. If there is any question about your safety, do not perform this step.

8. Call a qualified cleanup contractor if necessary.

9. Call the shipper or manufacturer's technical service representative.

10. Call the Chemical Transportation Emergency Center (CHEMTREC) at (800) 424-9300 or the EPA's Environmental Emergency Response Unit.

Emergency Response Steps Only Trained ERTs Should Take

OSHA's 29 CRF 1910.120 provides protection requirements for emergency response workers doing the following actions:

1. Set up command post and activate emergency response plan. Question transporter to determine composition and handling.

2. Put on protective clothing.

3. Remove injured individuals.

4. Check the shipping papers and package markings to determine composition and handling.

5. Dam up liquids if it can be done without injury.

6. Clean up spill.

7. Notify appropriate environmental agencies.

Emergency Response Guidebook

The Emergency Response Guidebook (which was suggested earlier) is provided by the DOT and is used by ER personnel when handling hazardous material incidents. It is composed of the following sections: numerical index of four-digit ID numbers, name of material in alphabetical order and response guides organized according to single or multiple hazards.

Personal Protective Equipment (PPE)

To prevent death or injury of themselves, the ER team must put on the proper protective equipment before entering the contaminated area. Information about the proper PPE can be obtained from OSHA, the National Institute for Occupational Safety and Health (NIOSH), the American Council of Government Industrial Hygienists (ACGIH) and the MSDS. The chemical may not even be known at this point; if this is the case, the equipment selection should be Level A protection. A variety of materials are used for the protective equipment, depending on the types of chemicals encountered. Butyl rubber, polyvinyl chloride and viton are three possible materials. The four levels of protective equipment, in decreasing order of protection, are A, B, C and D, and are illustrated in Figure 8-2.

Level A   Level A refers to maximum respiratory and eye and skin hazard situations. Level A is used if there is unknown hazard or strong toxic substances known, which require the highest level of protection. Level A protection also requires suitable rescue equipment present and the buddy system between workers. Examples of Level A equipment include self-contained breathing apparatus (SCBA), totally encapsulating vapor-tight suits with full-face piece, boots and gloves attached. A hard hat and two-way radio should also be worn inside the suit.

LEVEL A Protection
Totally encapsulating vapor-
tight suit with full facepiece
SCBA or supplied-air respirator.

LEVEL B Protection
Totally encapsulating suit
does not have to be vapor-tight.
Same level of respiratory
protection as Level A.

LEVEL C Protection
Full-face canister air
purifying respirator. Chemical
protective suit with full body
coverage.

LEVEL D Protection
Basic uniform, i.e.,
longsleeve coveralls, gloves,
hardhat, boots, faceshield,
or goggles.

Figure 8.2 Sample Protective Ensembles. (Reprinted from EPA, *Standard Operating Safety Guides*, 9285.1-03, Washington, 1992.)

Level B   Level B is selected when the type, atmospheric condition and concentration of the substance is known and has been determined to require maximum respiratory and high skin hazard protection. The highest degree of respiratory protection is required, but the toxics do not present a danger to small unprotected areas of the skin. Level B equipment includes SCBA and a two-piece, hooded and chemical-resistant suit. Outer and inner gloves, chemical resistent steel-toed boots and a hard hat are also recommended.

Level C   As with Level B, Level C is used when concentrations and types of hazardous substances are known and determined to require minimum respiratory and minimum skin hazard protection. Exposure of unprotected parts of the body is unlikely, and the monitored level of hazardous substance is 75% or less of the known Threshold Limit Value. Examples of Level C equipment include a half-face or full-face air-purifying respirator and a two-piece hooded chemical-resistant splash suit or clothing.

Level D   Level D equipment must also be used when the concentration and type of hazard is known. In this case there is no respiratory or skin hazard. There is no indication of airborne health hazards, or exposure of the body to hazardous materials is unlikely. Examples of Level D equipment include safety glasses, goggles, hard hat, face shield, gloves, chemically protective outer boots and dedicated work clothing.

Cleanup of Spills

Actions must be taken to clean up, package and dispose of the spill and to completely decontaminate the area. Once this is completed, sampling must be done.

A variety of absorbents, barriers, neutralizers and suction devices have been designed and are on the market for use in spill cleanup. Once the spill has been cleaned up, cleanup materials must be disposed of as a hazardous waste unless the spill is determined not to be hazardous. For example, if neutralizers bring the substance to an acceptable pH level (2-11) and no other contaminants are present, it would not have to be treated as hazardous waste.

Decontamination of Structures

If building walls, AC and heating systems and other structures are contaminated, they must also be cleaned before it is safe for occupants to return. This type of decontamination is often difficult and more costly than cleaning up a spill. It usually involves surface cleaning and may also require shallow and deep cleaning.

Surface cleaning is obviously the easiest to do and is either dry or wet removal. Dry removal includes vacuuming or removal of strippable coatings. PCB soot is often handled in this way. Wet removal includes washing with water, detergent and solvents and may require paint removal.

Shallow-depth decontamination is usually down to approximately one inch. It may involve sandblasting, shot blasting, scarifying, scabbing and high-pressure water treatment.

Deep decontamination is the most expensive process. This may involve demolition, excavation or application of special penetrants. Building and equipment decontamination is similar to spill cleanup. The first step is to make an assessment and perimeter definition. This is followed by cleanup standard identification and closure plan development. Notification, source removal and gross cleanup follows. Final decontamination, closure sampling and reporting conclude the process.

After the spill has been cleaned up and the HW placed in DOT-approved containers, normal hazardous waste disposal regulations would apply. There is, however, an abbreviated process for obtaining a quick, provisional, one-time EPA ID number for spills.

Reports

Reports must be submitted after spills occur. A spill incident report needs to be submitted within 15 days (40 CFR 263.30). This applies when the spill exceeds the RQ for the chemical and/or could potentially impact the environment or human health. One would consider the environment impacted if the chemical entered soil, water or the air in the RQ amount. Appendix A contains the federal list of hazardous substances and reportable quantities (40 CFR, Table 302.4). A copy of the manifest must be attached to an incident report along with an estimate of the type and quantity of the spilled wastes and the name of the recycle or disposal facility. Submitting the proper reports provides an opportunity for the agencies and responders to evaluate the response methods and procedures utilized.

Oil Spills

Some of the worst hazardous waste incidents in the last few years in terms of environmental damage have been oil spills. Oil spills can happen in several ways; however, ruptured tankers and refinery tanks and pipeline release have caused the most problems recently.

Recent Examples

On March 14, 1991, it was announced that the Exxon Valdez spill settlement had been set at over one $1 billion. This is one of the largest environmental settlements in history. Intensive cleanup activities lasted for at least two years (see Figure 8-3). Details of this incident are given in the case study at the end of this chapter.

Figure 8.3 Exxon Valdez. (Reprinted from EPA, "Protecting the Earth," *EPA Journal* 15/4, Washington, 1989.)

In February 1991, millions of gallons of oil were released from refinery pipelines in Kuwait. This massive and devastating release was associated with the Persian Gulf War. Because of the war, political reasons and other priorities (such as protecting drinking water systems), cleanup has been very slow. The amount of environmental damage from this event is immense and will be present for many years to come.

Type of Damage

Oil spills result in damage to many components of the ecosystem, especially the aquatic systems. Oil can lead to direct lethal toxicity to organisms. Sublethal disruption of physiological or behavioral activities is second in terms of devastation. This in turn reduces species resistance to infection or stress. Direct coating with oil impedes vital processes of respiration and feeding and prevents light penetration to plants. This coating

effect also increases temperature by absorbing solar radiation. Dermal toxicity to animals may result from direct coating with oil spills. Fish and
water fowl are especially effected, which has a significant impact on the
food chain.

Regulatory Aspects of Oil Spills

Two agencies are basically responsible for regulating oil spills. The
EPA sets regulations for inland spills. The Coast Guard is responsible for
spills in coastal waters and the Great Lakes. Many other organizations are
involved in terms of research and assistance, such as the National Oceano-
graphic and Atmospheric Agency (NOAA), the DOT, the Department of
Energy (DOE) and the Office of Research and Development in the EPA.

The Oil Pollution Act of 1990 is an example of some recent legislation. It establishes compliance standards for oil production, transportation
and storage industries. For example, a visible sheen on the water surface is
considered an RQ for reporting purposes as defined in 40 CFR. This
applies to state and federal navigable waters. The report must be made
within 24 hours. If this is done, it will free the person making the report
from criminal prosecution.

The EPA requires Spill Prevention Control and Countermeasure
Plans (SPCCP) of any facility that might spill into U.S. waters. This would
apply to spills from tanks and pipelines. The SPCCP generally applies to
anyone who stores 1320 gallons or more aboveground in combined containers or 660 gallons in any one container or 42,000 gallons or more below
the ground (40 CFR sec. 311). The SPCCP requires secondary containment
for all affected containers.

California Senate Bill 2040, which is the Lempert-Keene-Seastrand
Act (Government Code 670.1) and Management Review, was passed in
1990. This bill imposes many oil spill requirements and is like its Federal
counterpart, the Oil Pollution Control Act of 1990 (33 U.S.C. 2701).

Leaking Underground Storage Tanks

Fuels, solvents and various other chemicals are stored in underground storage tanks (USTs). Unfortunately, many USTs have leaked,
causing water and soil pollution. Almost every day, you can see another
contaminated site with mounds of dirt piled around the hole where a UST
use to be. Any tank with at least 10 percent of its volume beneath the sur-

face of the ground, including any pipes that are attached, meets the definition of a UST.

Some areas of the country have had so many leaking tanks, discharges and spills that the entire ground water system is riddled with contamination. Silicon Valley is a perfect example of this, with some areas hard to study and treat because of commingled plumes.

Regulatory Requirements

RCRA contains some UST regulations, especially the 1984 amendments. Leak detection methods, record maintenance, corrective actions, corrosion protection and final closure requirements are examples. Requirements can be found in 40 CFR, Subtitle C (hazardous waste) and Subtitle I (product).

In the early 1980s, HMW regulations, especially UST regulations, started to evolve in the San Francisco Bay area. The California Regional Water Quality Control Board sent out a survey in 1981 and many contaminated sites were identified. This resulted in the formation of the Santa Clara Toxics Coalition in 1982 and 1983 to address tanks. In 1983, Santa Clara County led the nation with the Model Hazardous Material Storage Ordinance and requirements for USTs. Also in 1983, the California Sher Bill imposed UST requirements and registrations, which evolved to other HMW in the Hazardous Substances and Waste Act. The 1983 Sher Bill was instrumental in establishing UST regulations across the United States.

In 1984, Congress adopted a national UST regulatory program, which is codified as Subchapter IX of RCRA. This Subtitle requires EPA to issue regulations establishing technical requirements for construction standards for new UST, leak detection and reporting. It also specifies practices, record-keeping, closure standards and financial responsibility.

Tanks installed after January 1, 1984, have numerous regulatory requirements, many more than existing USTs. For example, new USTs must provide a certificate of proper material and installation, primary and secondary containment, corrosion protection, spill and overfill prevention and continuous leak detection. There are some additional requirements for chemical USTs, such as secondary containment and interstitial monitoring.

Existing USTs have upgrade or replacement requirements. Corrosion protection, phase-in leak detection and spill and overfill prevention devices are required. Federal compliance for existing USTs was required by December 22, 1998, except for specific leak detection requirements.

If a leak occurs, whether a tank is new or old, the operator is responsible for reporting within 24 hours and then cleaning up. They must promptly notify the agencies and then take corrective action. This will probably require proper tank closure. In addition, the owner or operator of all USTs will have to provide financial assurance for taking corrective action and compensation for injury or damage before leaks occur.

When a tank is taken out of service it must be decomissioned properly. For example, the tank must be filled with a solid inert material or a foam. Foams are gaining popularity because a foam-filled UST can be removed intact for disposal.

Line leaks associated with tanks have been addressed by two recent regulations. Pressurized piping must be equipped with line leak detectors, and testing conducted, as of December 22, 1990. As of September 22, 1991, UST operators were to meet new automatic line leak detection requirements on new or existing underground pressurized piping. Details concerning leak detection in lines can be found in 40 CFR 280.44.

Major Causes of Leaks

The greatest cause of UST leaks is lack of corrosion protection. Steel can be protected by several forms of cathodic protection: sacrificial anodes, impressed current and corrosion protection coating (such as galvanization or zinc coating). 40 CFR 280 presents the operation and maintenance of corrosion protection systems.

A second cause of leaks is improper installation. This can be avoided by following installation procedures designed to minimize leaks and other problems.

A last major cause of leaks occurs during tank filling when spills and overfills occur. By following the correct tank filling practices, such as use of overfill alarms, this type of accident can be reduced. 40 CFR 280.30 delineates spill and overfill control.

Leak Detection Methods

Several methods are available for leak detection. Whatever the method utilized, records must be kept as prescribed by the regulations. Operators are required to do leak detection, which includes annual tank tightness testing, monthly monitoring of the product level and inventory control. This may include a tank receiving and dispensing "balance sheet." In special situations, the following additional leak detection may also be necessary: monitoring for vapors in the soil, monitoring for contaminants in the ground water, monitoring an interception barrier and interstitial monitoring within secondary containment. 40 CFR 280.43 lists methods of release detection.

Cleanup of Leaks and Site Contamination

If a tank, pond, pipeline, land disposal site or other source releases an untreated toxic chemical into the environment for a "longer" (generally more than just an instantaneous spill) period of time, considerable damage can occur, especially to the soil and ground water. Cleanup usually becomes significantly more involved than that for the typical highway-type spill. Cleanup in these situations is usually regulated in one or more of the four following ways: state-hazardous-waste-act-regulated; RCRA-regulated; state-Superfund-regulated; or Federal-Superfund-regulated. Cleanup of these longer term, more involved sites usually involves containment systems and/or treatment systems.

Containment

Containment is usually only a short-term solution or fix. This is due to the fact that most containment structures will eventually leak. There are many different types of containment due to the variety of real-life situations. The preferred structure in one environment would probably not be suitable in most other situations.

The major types of ground water barriers or containment systems include slurry walls, grout curtains, vibrated beams (enhanced slurry wall), sheet piles and block displacement (placing a barrier beneath the surface). All of these systems minimize the off-site movement of the contaminant by reducing ground water flow. The barriers need to be tied into a naturally occurring geological layer of low permeability to be effective. Certain other control systems could loosely be considered containment as well, such as ground water pumping, subsurface drains, runoff controls, surface seals and caps, solidification, stabilization and encapsulation [Wentz 1989]. Many of these containment systems are used together, along with several of the treatment systems presented below. By combining systems, the cleanup can be made more cost effective.

Treatment

Treatment is also presented in detail in Chapter 11, with an emphasis on off-site treatment of normally generated hazardous waste. The treatment discussed here will stress on-site treatment of spills. Figure 8-4 presents the major types of treatment selected for National Priority List (NPL) sites in fiscal year 1991. These are categorized into innovative technologies vs. established technologies. The rest of the discussion in this section will concentrate on the major innovative approach, *in situ* treatment, and the predominant established technology, off-site treatment.

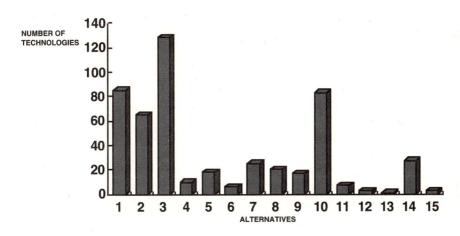

ALTERNATIVE KEY

Established Technologies (228) 58%
1 Off-site incineration (85) 17%
2 On-site incineration (65) 13%
3 Solidification/stabilization (128) 26%
4 Other established (10) 2% —this includes soil aeration, *in situ* flaming and chemical neutralization

Innovative Technologies (210) 42%
5 Soil washing (18) 4%
6 Solvent extraction (6) 1%
7 Ex situ bioremediation (25) 5%
8 *In situ* bioremediation (20) 4%
9 *In situ* flushing (17) 3%
10 Soil vapor extraction (83) 17%
11 Dechlorination (7) 1%
12 *In situ* vitrification (3) < 1%
13 Chemical treatment (1) < 1%
14 Thermal desorption (27) 5%
15 Other innovative (3) < 1%—this includes air sparging and contained recovery of oily wastes

Figure 8.4 Alternative Treatment Technologies Selected for NPL Sites Through Fiscal Year 1991 (total number of technologies = 498). (From EPA/TIO, January 6, 1993.)

In situ treatment   In-place- or *in-situ*-treatment is preferred by many agencies since it minimizes the amount of waste that must be transported and disposed of off-site. *In situ* treatment methods for contaminated soil and ground water include vapor extraction, steam stripping, thermal desorption, chemical injection, detergent wash and treatment, passive biodegradation, enhanced biostimulation, flushing and vitrification. As Figure 8-4 shows, soil vapor extraction is done more than any other *in situ* innovative technology (17%). Examples of some of these technologies are shown schematically in Figure 8-5.

In situ steam stripping has been fairly successful for certain types of contamination. This process basically starts with steam being generated (center of the first part of Figure 8-5) and then injected down a well (left side). This action gets the contaminants, such as volatile organic compounds (VOCs), to off-gas. The off-gas is collected in a shroud, which is then transferred via suction to a scrubbing system. From here a condensate forms, after gas cooling, which contains the contaminants. The rest of the system is basically in support of these key components. This process is effective at hydrocarbon contamination levels of up to 20,000 to 30,000 ppm. The cleaned soil will contain low ppm levels.

In vacuum extraction, or soil ventilation, a vacuum is applied to a well. Volatile contaminants then move out of the soil. This method has the drawback that it works for only volatile compounds and it moves the contaminant from one medium (soil) to another (air). Vacuum extraction has been used mostly for treating gasoline- or solvent-contaminated soils, which are relatively permeable.

Thermal desorption is used occasionally to remove volatile organics from soil. In this process warm air is fed into the waste deposit. This causes the volatile organic compounds to move into the gas phase, which can then be treated.

Immobilization and *in situ* vitrification are not as common; however, in certain situations they may be the perfect solution. Immobilization makes use of asphalt or the Portland Cement Pozzolan Process. Cement is mixed with the hazardous waste and allowed to solidify, which minimizes leachate out of the concrete material. This process is very similar in outcome to *in situ* vitrification, which subjects soil to extreme heat and results in a glassy matrix. *In situ* vitrification may cause some contaminant organics to vaporize and escape for the matrix, which in turn would have to be collected by a vacuum system.

Process flow diagram of an in situ steam stripping process.

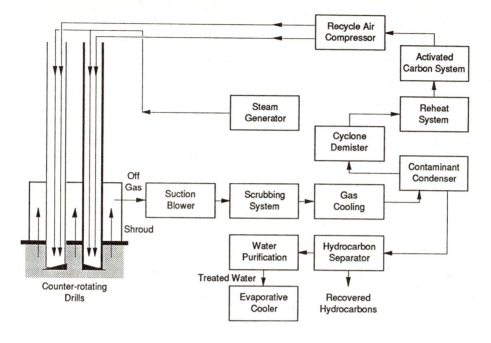

Pilot-plant flow diagram of an in situ aerobic/anaerobic biodegradation system.

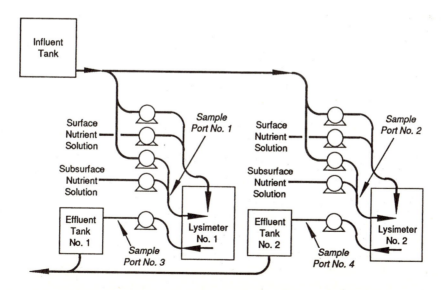

Figure 8.5 *In Situ* Treatement of Hazardous Waste Contaminated Soils (Part 1 of 2). (Reprinted from EPA, *Handbook on In Situ Treatment of Hazardous Waste-Contamination Soils,* EPA Document No. 540/2-90/002, Ohio, 1990.)

Process diagram of an in situ vacuum extraction process.

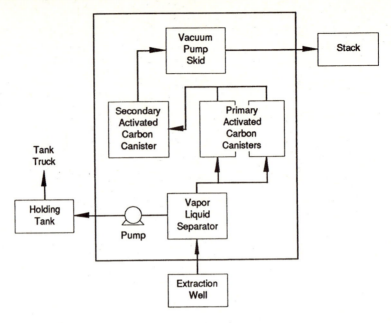

Schematic of a biostimulation bioremediation system.

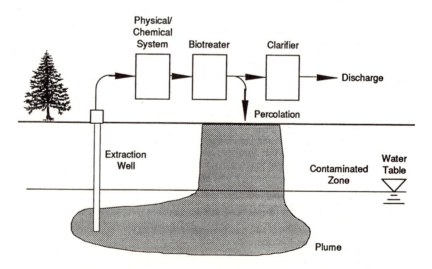

Figure 8.5 *In Situ* Treatement of Hazardous Waste Contaminated Soils (Part 2 of 2). (Reprinted from EPA, *Handbook on In Situ Treatment of Hazardous Waste-Contamination Soils*, EPA Document No. 540/2-90/002, Ohio, 1990.)

Of all the methods mentioned, bioremediation and biodegradation seem to be getting the most attention. Biodegradation involves the use of natural or introduced microorganisms, such as bacteria. Growth of the microorganisms is stimulated, and as they grow, they utilize certain organic contaminants as a source of food in their metabolic process. To be effective, the moisture, nutrient, pH and oxygen contents of the soil must be maintained at prescribed levels. In many cases, but not all, the metabolic by-products are nontoxic (carbon dioxide and water). The result is that the problem has been resolved, and the land is reusable without the need to involve other locations.

Removal and treatment of soil If the above *in situ* treatment is not possible, it may be necessary to remove the soil and treat it on the surface or off-site. Once it is removed, a variety of options are available. For example, biodegradation and land treatment can occur off-site and the soil returned. Depending on the contaminant some other form of stabilization or even solidification may be necessary. A last resort would be excavation and disposal off-site, especially for small areas of contamination.

Bioreactors have been designed to be used in certain biodegradation options. Contaminated soil is placed in a container along with charcoal or diatomaceous earth, nutrients and the bacteria. Air and the right amount of water are also needed.

In certain cases mobile incinerators are used for short-term on-site or off-site incineration. The soil is heated to vaporize the contaminants, with the exhaust gas usually reaching 2000 to 2400 degrees F. Mobile incinerators can effectively treat soil with nearly any type of hydrocarbon contamination. For this to occur, the incinerator must be a permitted transportable treatment unit (TTU).

Removal and treatment of ground water Once the soil treatment is completed or well underway, the ground water treatment should start. As with soil, if the ground water can't be treated *in situ*, it is necessary to bring it to the surface and then treat it on or off-site. Ground water is removed by extraction wells, treated to specified levels, discharged to a POTW or surface water body or returned via injection wells. Common treatment processes include oil-water separation, air stripping, carbon adsorption, ion exchange, solidification, filtration, sedimentation, centrifugation, flotation, distillation, evaporation, flocculation, oxidation and chlorination.

Ground water treatment technologies can be categorized in many ways. For example, they can be divided into inorganic and organic technologies. Examples of inorganic component separation include ion exchange, carbon adsorption, reverse osmosis, electrodialysis and crystallization. Examples of organics component separation include resin adsorption, steam stripping, solvent extraction and distillation.

Treatment can also be discussed in terms of chemical or biological treatment of ground water. Chemical treatment includes neutralization, precipitation, ozonation, reduction, calcination, electrolysis, photolysis and others. Biological treatment techniques include the use of activated sludge, aerated lagoons, trickling filters, anaerobic digestion and composting.

Oil-water separation is used in many cases. Water with the oil is pumped into a tank where the oil floats to the top and is skimmed off. Water without oil is then returned via an injection well. Internal baffles, parallel inclined plates or plastic coalescers are often added to assist in this process. The effectiveness of the separation process can also be influenced by the waste stream's rate of flow, temperature and pH.

Air stripping involves pumping contaminated water (of certain types) into the top of a tower that is filled with whiffle balls. The water trickles to the bottom of the tower and the contaminants volatilize out of the water. As with vapor extraction, air pollution may become an issue; therefore, air emission recapture and treatment should be considered. Air stripping is good for treating highly volatile low-water-soluble aqueous organic wastes, such as chlorinated hydrocarbons, tetrachloroethylene and aromatics.

Carbon adsorption is also commonly used. Contaminates are adsorbed onto the carbon surface. This can be very expensive, in terms of regeneration and disposal. This process is used to treat single-phase aqueous organic wastes with high molecular weight and low solubility and polarity. Chlorinated hydrocarbons, suspended solids of less than 50 ppm and dissolved inorganics of less than 10 ppm are treated using carbon adsorption.

Regulatory Aspects of Cleanup

Cleanup of the non-emergency type spill is usually handled in one or more of the following ways: RCRA-regulated, CERCLA/SARA-regulated, state-Superfund-regulated, state-hazardous-waste-act-regulated and through private action without involvement of the federal or state regulatory agencies.

RCRA-Regulated Cleanup or Corrective Action Process

The RCRA-regulated cleanup process is illustrated in Figure 8-6. The final RCRA Corrective Action Program rule was adopted in January

1993 in 40 CFR 261, Subpart S. This type of cleanup is imposed by permit or enforcement order; it does not involve Superfund. As can be seen from this figure, there are primarily four phases: RCRA Facility Assessment (RFA), RCRA Facility Investigation (RFI), Corrective Measure Study (CMS) and Corrective Measure Implementation (CMI). Some of these phases can overlap and, in rare cases, even run in parallel. They are designed to provide safety to people living near the site and to clean up personnel. They are also designed to achieve the maximum cleanup of the environment in a cost effective way. With regulatory approval, pilot plants or some initial cleanup can occur during early stages, such as during the RFI. Doing this not only tests out certain technologies on a smaller, less costly basis, but also starts to control the problem. The CMI period is shown in year six. It is important to note that the figure means the start of CMI. One site the author was involved with had a 30-year CMI.

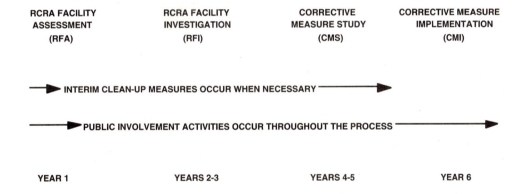

Figure 8.6 The Corrective Action Process. While some of these steps are time consuming, they may occur simultaneously. Each step is essential to ensure efficient use of funds and the safety of people living near the facility. The EPA recognizes the threat posed by environmental contamination, the need to implement the cleanup program expeditiously, and the importance of keeping the affected community informed. (Reprinted from EPA, *Cleaning Up Hazardous Waste Sites,* San Francisco.)

Comprehensive Environmental Response, Compensation and Liabilities Act (CERCLA) Regulated Cleanup

While RCRA is oriented more toward preventing future hazardous waste problems, CERCLA's and SARA's emphasis is more toward the cleanup of past hazardous waste situations. A past or historical problem or release could even mean ten minutes ago, especially if the responsible party just went bankrupt.

The Comprehensive Environmental Response, Compensation and Liability Act (CERCLA) was passed in 1980 and gives EPA and states authority and a fund to respond to uncontrolled problems at hazardous waste disposal sites and almost any other site contaminated with hazardous waste. The act can be found in 42 U.S.C. 9601-9675 with regulations in 40 CFR. Superfund is the heart of CERCLA and is a collection of money from special taxes. This money is used to clean up sites if private funds are not immediately available.

The Superfund process is very involved, as Figure 8-7 illustrates. The National Priorities List (NPL) is shown as the third item of the figure and is a key component of the overall system. The NPL is the list of sites that EPA focuses on for cleanup under CERCLA.

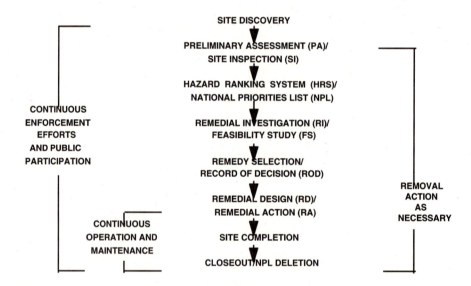

Figure 8.7 The Superfund Process. (Reprinted from EPA, *CERCLA/Superfund Orientation Manual,* Washington.)

CERCLA establishes liability for the cleanup of active or abandoned sites causing environmental damage. A generator can be held responsible for remedial or cleanup costs if their waste is found at a Superfund site.

The courts, in interpreting CERCLA, determined that this liability is strict, joint and several. This has had a monumental impact on all segments of society. Strict liability basically means that if it is your waste you are liable, even if you were not negligent. Joint liability could make you totally responsible even if you have only limited involvement. For example, if 1% of the waste is yours and the others involved are broke, you pay 100% of the cleanup. Several liability requires you to clean it all up, if the others won't, if you have any liability at all. It's up to you then to try to get compensation from the other potentially responsible parties (PRPs).

The overall goal of Superfund is to derive maximum benefit for the nation as a whole, not just one site or state. For example, excavation in one state and disposal in another state is to be avoided.

National Priorities List (NPL)   As previously mentioned, the NPL is a listing of sites identified by the EPA for Superfund. The sites are arranged according to a Hazard Ranking System (HRS) score. The HRS score is a standard set of factors related to migration of contaminants in the environment. The probability of contamination via three routes is considered: ground water, surface water and air. An assessment of explosiveness, flammability and potential for direct human contact is also considered in the HRS. HRS scores range from 0 to 100. A score of 100 would be the most hazardous site. It usually takes a score of greater than 20 to 30 to place a site on the NPL. As Figure 8-8 shows, there are several thousand NPL sites in different stages of the process. The most important figures are shown in the last two bars; that is 149 sites completed and 465 remedial actions started as of November 11, 1992. These numbers reflect actual progress in the field that has, or is presently underway to address the problems.

National Contingency Plan (NCP)   The   NCP   comprises guidelines that the EPA must follow in implementing Superfund and private parties must follow if they are to be able to seek reimbursement from other PRPs under Superfund. The NCP has two general categories, removal actions and remedial actions. Removal actions must be completed within six months and are specified if a prompt response to prevent harm is necessary. Removal of waste, fencing and barriers are examples of this type of action. Removal actions occur when an urgent response is necessary to minimize significant damage to health or the environment, such as a leaking derailed train car. EPA can also initiate removal actions at non-NPL sites.

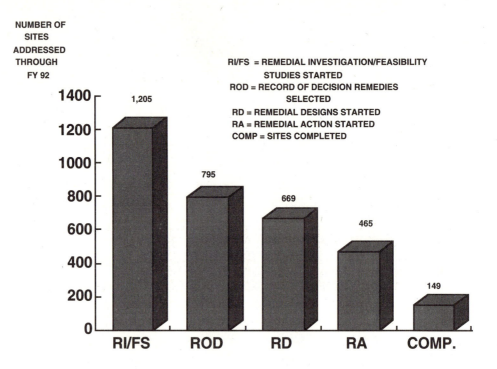

Figure 8.8 Work has begun at most NPL sites. (From EPA, *Superfund Enforcement Speaker's Kit*, CERCLIS, Washington, 1993.)

Remedial actions are used when there is not an emergency or following removal actions. Remedial actions are longer-term, more expensive, permanent remedies. Clay caps and alternate water supplies are examples.

Federal, state and private contributions As of November 11, 1992, private parties were paying for approximately 72% of the Superfund cleanup costs, as illustrated in Figure 8-9. This is a 30% increase since SARA was passed. Much of this comes from the "Deep Pocket," or financially strong organization, who may only be indirectly responsible in some cases. This includes costs that the EPA has paid and then recovered from the private party. As of fiscal year 1992 the EPA reported that the cumulative value of Superfund settlements totaled $842 million. This amount of money was made up of 1355 settlements for cost recovery.

The 28% not covered by private parties must be covered by the Superfund trust fund. Superfund is especially important for abandoned

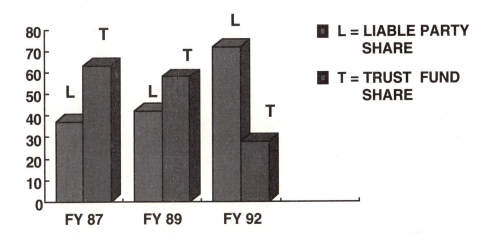

Figure 8.9 Payment of Superfund Costs: Remedial Action Starts. (From EPA, CERCLIS, November 11, 1992.)

sites since it is sometimes hard to find responsible parties with enough money to fund the cleanup.

Stages of Superfund investigation (NPL sites)

1. Remedial investigation/feasibility study (RI/FS)—The goal of the remedial investigation phase is to characterize the site by type and extent of contamination. The feasibility study will discuss the issues and risks involved, the proposed remedial alternatives and the potential environmental impact of each.

2. Record of decision (ROD)—The ROD phase documents the action plan chosen for remediating a site. The ROD also provides a basis for cost recovery from unwilling responsible parties.

3. Remedial design (RD)—The objective of this third phase is to prepare the design plans and specifications for performing the cleanup.

4. Remedial action (RA)—The remedial action is the last phase and is the actual cleanup operation.

Superfund site cleanup from differing perspectives The mention of Superfund involvement strikes terror in the hearts of most organizations. Many people are unsure how to handle such a situation. Often it is not clear who is responsible and to what extent. The issue of "how clean is clean" can also cause considerable confusion and uncertainty.

Early involvement by an organization can have both positive and negative aspects. On the positive side, it is attacking the problem early before it gets out of hand. On the negative side government agencies often require more of those organizations that step forward, since they are readily available. Either way, it is important for any PRP to quickly collect as much information as possible and to try to understand the situation. It is also a good idea to bring in all PRPs to share in the expense.

Superfund can also cause problems for agencies. Paperwork nightmares and litigation can quickly become a drain on the agency's resources.

Superfund accelerated cleanup model (SACM)   SACM was introduced in December 1992 in order to make Superfund cleanups more timely and efficient. This in turn will reduce impacts on the environment. One component of SACM is early removal of contaminated materials while long-term cleanups are in design. A regional decision team will quickly decide if a site requires early action, long-term action or both. Other early actions could include access restriction and measures to keep contaminants from moving off-site (Figure 8-10).

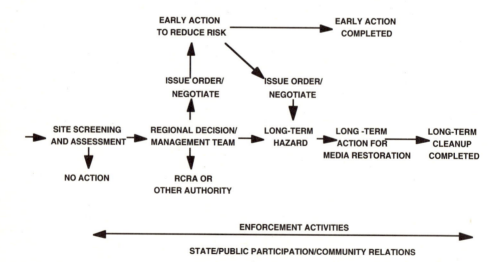

Figure 8.10 The SACM Process. (Reprinted from EPA, *CERLCA/Superfund Orientation Manual*, Washington.)

Superfund Amendments and Reauthorization Act (SARA)

SARA was passed in 1986 to extend CERCLA. In addition to $8.6 billion in funding, SARA requires more stringent standards, which increase cleanup costs. SARA also establishes enhanced public participation, a site investigation, feasibility study and remedial action schedule for EPA.

Case Studies

Yusho, Japan

In 1968 in Yusho, Japan, a heat exchanger sprung a leak and contaminated a large quantity of rice oil. More than 1000 people used the contaminated oil and developed a wide variety of complaints, including abdominal pain, fatigue, coughs, eye discharges and nervous system disorders. The heat exchanger contained PCBs. As a result of the health data, in 1972 the Japanese government banned the production, exportation and importation of PCBs [Allegri 1986].

Biocraft Laboratories—Waldwick, New Jersey

Biocraft Laboratories opened in 1972 to manufacture synthetic penicillin. Their operation included ten 10,000-gallon underground storage tanks containing n-butyl alcohol, acetone, methylene chloride, wastewater and spent solvents.

Sometime between 1972 and 1975, two pipes leading to the tanks started leaking and contaminated the soil and ground water. Degradation of a nearby brook led the New Jersey agency to require tank and pipe testing. The agency estimated the leak to exceed 33,000 gallons. Site investigation, groundwater monitoring and treatment were performed [Wentz 1989].

San Gabriel Basin

One example of a Superfund site and the associated ground water contamination is shown in Figure 8-11. This illustrates that the extent of groundwater contamination can be extremely great in some cases, such as in the San Gabriel Basin. This complicated Superfund cleanup is one of EPA's largest. It involves approximately 200 square miles of aquifers near Los Angeles [EPA 1989].

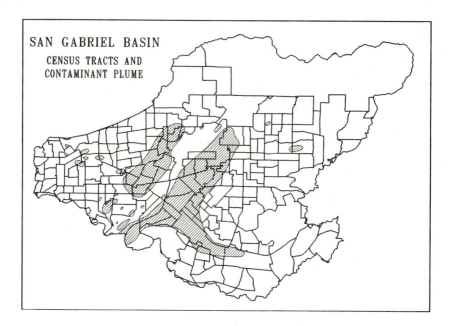

Figure 8.11 (Reprinted from EPA, *Environmental Data: Providing Answers or Raising Questions.* EPA Journal No. 15/3, Washington, 1989.)

Exxon Valdez

On March 24, 1989, the Exxon Valdez hit Bligh Reef in Prince William Sound, which is a port off the Gulf of Alaska. Eleven million gallons of crude oil poured out of the ruptured hull. Within two or three weeks, the oil had spread as far as five-hundred miles and involved numerous islands and sensitive wildlife areas.

This event led to the death of at least 100,000 sea birds, including 160 bald eagles, 1000 sea otters and unknown numbers of salmon, herring and halibut. It also resulted in over one hundred and fifty lawsuits and $1 billion expense to Exxon.

Most important, over 11,000 individuals have worked to clean up the spill. As of this date, progress has been made; some areas are starting to recover [EPA 1989].

CHAPTER

NINE

Use and Storage of Hazardous Materials and Waste

Introduction

This chapter is the beginning of the discussion of non-spill-type HMW controls and management systems that follow the life cycle of an HMW presented in Figure 1-2. These are the procedural systems that everyone involved with HMW should implement to minimize exposure and risk. In many cases, these systems are mandated by law. However, proactive and innovative management requires implementation of systems that have not yet been required by regulation. Many of these practices will be needed to meet future ISO certification, or, in other words, to sell the product or service. Most of these practices are commonly called total environmental quality management (TEQM) and are an offshoot of the total quality management approach. A strong proponent of TEQM is the Global Environmental Management Institute in Washington. TEQM advocates error-free work and preventing problems before they occur [Meehan 1993]. The rest of this book presents TEQM systems and management controls that will help achieve ISO certification and regulatory compliance.

The field emphasized in Chapter 9 is industrial hygiene. Industrial hygiene is a science and art devoted to the anticipation, recognition, evaluation and control of employee exposures. The exposures are environmental factors or stresses, arising in or from the workplace.

The exposures can cause sickness, impaired health and well-being or significant discomfort and inefficiency among workers.

Worker safety concerning the use and storage of HMW also varies, depending on the area of the world. In the author's estimation the U.S. and Europe worker safety controls are roughly comparable. The Far East again falls behind in terms of safety regulations, awareness and controls concerning use and storage. For example, in Thailand's electronics industry it is extremely difficult to get workers to even wear safety glasses. Therefore, injuries due to chemical splash in the eye are commonplace.

Before a discussion of worker safety controls can begin, some definitions concerning exposure levels will be presented. Time-weighted average (TWA), or the airborne concentration of a material to which a person is exposed, averaged over time, is one way of expressing employee exposure to an HMW. TWAs assume an eight-hour exposure over a 40-hour work week. Closely related to TWAs are ceiling or maximum short-duration values.

Employee handling and use of HMW must meet OSHA's Permissible Exposure Limits (PELs). The PELs are calculated as a TWA. The PELs define the amount of hazardous airborne chemicals to which an employee may be exposed over specific periods of time. Besides long-term eight-hour exposure, PELs also present short-term exposure limits for 15-minute time periods.

Employee handling and use of HMW should also meet specifications of the Association of Certified Government Industrial Hygienists (ACGIH). This organization uses threshold limit values (TLVs). TLVs are also expressed as a TWA. In many cases a PEL is the same as a TLV. In general, all exposure limits and recommendations are set to protect the average worker. Certain workers, however, may be more sensitive to various chemicals or combinations of chemicals. In these situations, the worker safety controls that will be discussed next must be implemented earlier than what the published limits and recommendations specify.

Worker Safety Controls

In industrial hygiene and many other fields as well, there is a control hierarchy. The hierarchy considers built-in protection in the design of a process being preferable to any method that depends on continued human implementation. Therefore, engineering controls or built-in protection are most preferred, followed by administrative controls and, last, personal protective equipment (PPE).

There are direct and indirect exposures. Direct exposure does not involve transport media, such as air. A direct exposure example would be exposure to a drug. Indirect exposure involves environmental media transport, such as exposure to air pollution, surface and ground water contamination.

The routes by which toxic chemicals can enter the body are key factors in the proper determination of all worker safety controls. These exposure pathways include ingestion, inhalation, injection and dermal absorption. With ingestion, chemicals enter the body via food, water or objects placed in the mouth. The chemical then comes in contact with the cells in the gastrointestinal tract. Controls designed to protect inhalation hazards will normally protect the worker from ingestion of chemicals. It's during rest periods or lunch breaks that one worries more about ingestion hazards, especially if the break is taken in a laboratory, for example. Therefore, rest periods or lunch breaks should be limited to designated areas.

The inhalation route is a much greater concern and usually requires engineering, administrative and PPE controls. In general, inhalation brings the contaminant in contact with lung cells. Most of these chemicals are in gaseous form (formaldehyde), vapor form or very small particulates (coal dust). Inhalation is of special concern since the exposed surface area (lungs) is large and the lung is anatomically designed to allow rapid absorption.

Dermal absorption involves the movement of chemicals through the skin or eyes. The amount of absorption is largely dependent on the condition of the epidermal layer of cells and the chemical. For example, if the skin layer is not scratched or broken, absorption will be much less. The skin is anatomically designed to make absorption of chemicals much more difficult than that which occurs in the GI system and lungs.

Injection is the fourth route of entry. It involves injection through the epidermis and may occur when handling sharps, compressed air, gases or other materials.

Engineering Controls

Engineering controls are the top priority in terms of HMW use and storage. For engineering controls to be successful they must be incorporated into the design specifications. Engineering controls concentrate on substitution, isolation, enclosure (shielding from employees), wet methods and local exhaust ventilation.

Substitution is a very common type of engineering control. This could include process substitutes (efficiency improvements) and substitution of equipment (newer is usually better). Material substitution also occurs, that is, the use of a less hazardous product. Many of these actions could also be considered waste minimization or pollution prevention and are also discussed in Chapter 14.

Isolation, in time and distance, is another engineering control. The isolation can be of the material, equipment, process or operator; for example, relocating the work area so that fewer workers and neighboring residents are affected by the exposure. Another example is the use of robotics, which are commonly used in the electronics industry. This reduces employee exposure by reducing the number of employees.

Enclosure is an engineering control that can involve the entire process, certain pieces of equipment or only the operator. For example, with enclosure and negative pressure any leaks are to the inside of the enclosure, not outside. On the other hand, car toll-taker controls utilize positive pressure so that exhaust into the booth is reduced.

Wet methods of engineering control are usually required for asbestos, particle board manufacturing, mining and other dusty operations such as demolition and road repairs. This could include dust suppression water spray systems. When asbestos is removed, water sprays help keep the amount of airborne fibers at an acceptable level.

Housekeeping could be considered both an engineering control and an administrative control. This may include secondary collection of particles. For example, when materials settle out on surfaces, they should be cleaned up. Housekeeping may also involve containment of liquids and dry materials, such as with vacuum systems. Good housekeeping might include not mixing nonhazardous wastes with hazardous. If one puts nonhazardous cleaning agents or rags in the same container as a hazardous solvent, the entire contents becomes a hazardous waste. Mixing of different hazardous wastes is also to be avoided because of possible incompatibility. In addition, mixing may make recycling very difficult or disposal more expensive. Good housekeeping also includes minimization of spills, leaks and leftover products in containers. Frequent employee meetings to emphasize all the above is also a good housekeeping policy.

By far the most widely used engineering control for inside operations is local exhaust ventilation. This involves enclosing the source as completely as practicable and may also utilize positive and negative pressure. The better the enclosure, the less capture velocity needed, such as when a hood is used.

It is important to capture the contaminant with adequate velocities. It is also important to keep contaminants out of workers' breathing zones. This means pulling contaminants away from the operator. The system must supply adequate make-up air. If it doesn't, contaminants can't be pulled out of the area as well. Last, one must discharge the exhaust away from other intakes (such as HVAC) or through an air cleaning system.

Administrative Controls

Administrative controls are interim control measures, not final as were the engineering controls. They may involve temporary isolation (not permanent as with engineering controls) of the operation in time (such as at night). More commonly, the administrative controls limit the exposure time to reduce the concentration. For example, by rotating employees, exposure time can be kept at an acceptable level.

Proper work planning can also result in the minimum amount of time required for the performance of the job and therefore less exposure. Procedures, training and audits could also be considered administrative controls. For administrative controls to be effective they must be well documented and posted, and training must be provided. The training is of great importance.

Personal Protective Equipment

Personal protective equipment for day-to-day normal work operations (not including spill cleanup protection) is the last resort in minimizing exposures. It does not control the external environment and should be offered to employees only for added, nonrequired protection. Unfortunately, many employees will not utilize personal protective equipment for day-to-day operations, whether it is required or voluntary.

Examples of personal protective equipment commonly found in electronics manufacturing include ear plugs and muffs, safety glasses, hard hats, goggles, coveralls, masks, steel-toed boots, gloves, respirators and aprons. In this industry, the PPE and ventilation systems are as important in protecting the product as in protecting the worker.

One of the most common types of personal protective equipment is the respirator. Respirators are of the air-purifying or air-supply type. They can be disposable or reusable, and half-face or full-face. Figure 9-1 illustrates a worker in Thailand involved with paint spraying. On the left side the worker is not wearing a respirator, and on the right he is wearing one; however, the wrong kind. The illustration should show the chemical-cartridge-type respirator in place of a cotton mask. The most commonly used respirator is the chemical cartridge type. The cartridge contains various chemical substances that purify inhaled air of certain gases and vapors. Air-supply respirators provide better protection; however, mobility is impaired by either the supply line or tanks.

- การพ่นสี-ต้องสวมหน้ากาก และชุดทำงาน

☒ ไม่ ละเลยการป้องกันตนเองในขณะพ่นสี เพราะในสีพ่นมีสารพิษหลายชนิด อาจผ่านเข้าทางเดินหายใจ และดูดซึมผ่านผิวหนังทำอันตรายต่อร่างกายได้

☑ ต้อง สวมหน้ากากและชุดทำงานให้เรียบร้อยเมื่อต้องทำการพ่นสี

Figure 9.1 Personal Protective Equipment in Thailand.

Various regulations, codes and standards apply to respiratory protection. The most common include OSHA's 29 CFR 1910.134 and NIOSH—Guide to Industrial Respiratory Protection [Technical Report 1976].

Prior to respirator usage, there are some up-front actions that must occur. Inhalation and fit tests must be done on the personnel who are to wear the respirators. Some individuals can't wear respirators because of existing respiratory or other medical problems. Facial features must also be considered. For example, beards, and in some cases mustaches, must be removed because they prevent a good seal. Then it is a matter of picking the proper respirator and appropriate cartridges.

Employees must be trained in the use of PPE. This training should include equipment operation, maintenance, decontamination and emergency procedures. For example, if the equipment fails, the person needs to know self-rescue.

Many other factors should be considered when deciding the appropriate type of PPE. The potential type and level of contaminant are of greatest importance. Also important is the equipment's permeability, flammability, degradation, heat concentration, visibility and decontamination characteristics. Any piece of protective equipment has other limitations, such as worker acceptance and human error.

Storage

Introduction

As discussed in Chapter 2, the bulk of the regulations covering storage can be found in RCRA section 3004, 49 CFR, NFC, SARA and OSHA (29 CFR 1910). In California most of the storage regulations closely pattern federal ones and are found in the HWCL (Health and Safety Code 25123).

It is interesting to note that in many industries, such as electronics manufacturing, the phase-out of some chemicals, like Class I ozone-depleting substances (ODS), has resulted in the need to store much larger volumes of flammable chemicals, such as acetone and alcohol. Additional storage and dispensing areas are needed along with flammable cabinets, safety cans and increased ventilation.

In general, storage of HMW should be done in a way that minimizes accidental release. Various storage considerations are summarized in Table 9-1 and discussed in the rest of this section. The most important provisions include signs and labels, emergency capability/contingency plans, hazard communication plans, appropriate containers, physical design and duration of storage. Employee training in terms of safe handling is also an important consideration.

Storage Warning Signs and Labels

Usually, the manufacturer will provide MSDS and proper labels on their containers. If the worker uses that labeled container, a new label is not usually required. If, however, the chemical is transferred to smaller containers, these must be labeled with safety information.

An example of a chemical storage sign used in Portugal is shown in Figure 9-2. This sign is used for HMW. Use of the familiar skull and crossbones can be found in many countries.

Table 9.1 Requirements Concerning Aboveground Storage of Most Hazardous Materials

Requirements	Texas		Tennessee		California	Federal
	State	Local	State	Local		
Labeling	Yes (Board of Health)	Yes (City ordinance) County adheres to state regulations	Yes (TOSHA) (Hazard Comm. Std. Act)	Yes (Same requirements as TOSHA and OSHA)	Yes (Cal OSHA and DHS—Prop. 65)	Yes (Fed. OSHA and NFPA)
Secondary Containment	No	Yes (City ordinance) County adheres to state regulations	No	"Strongly encouraged"	Yes (Cal OSHA and Fire Marshall)	No (except by EPA for petroleum)
Contingency Plans	Yes (Model after other states or federal)	Proposed for Fall 1990 (City ordinance) County adheres to state regulations	Yes (1910120-L Haz. Waste Emergency Response and 191038 Emergency Response Evacuation Plans)	HMMP "strongly encouraged" but not required Tier 2 Report required	Yes (HMMP, HMBP, RMPP required by state and local agencies)	No (except EPA requires an SPCCP for petroleum)
Hazard Communication Plan	No	Yes	Yes (TOSHA)	Must be in compliance with TOSHA and OSHA	Yes (OSHA and Local)	Yes (OSHA)
Inventory	Yes (Department of Health)	Yes	No	No	Yes (DES and Local)	Yes (EPA and SARA III)
Miscellaneous Requirements	Training required		Employee training program required			NFPA CODE 49—Hazardous Chemical storage requirements

PERIGO

RESIDUOS

Figure 9.2 Portuguese Hazardous Waste Storage Sign.

It is also not uncommon to find unlabeled HMW stored in homes and industry. Obviously, this is extremely dangerous. The EPA, National Fire Protection Association (NFPA) and OSHA have all been working to correct this problem. The use of proper labeling minimizes the mixing of HMW and enhances good housekeeping.

EPA warning signs and labels   HW storage containers must be labeled with the words "hazardous waste" and the date on which each period of accumulation begins. For generators of less than 100 kg/month of hazardous waste, the accumulation date on the label should be the date when 100 kg (approximately one-half of a 55-gallon drum) of hazardous waste is reached, which is when the 90-day clock begins, but no longer than one year.

The words Hazardous Waste must be marked clearly on each container with hazardous waste. The chemical name, the physical state (solid, liquid, gas), hazardous properties of the waste (flammable, corrosive, etc.), composition, and the EPA ID number, name and address of the generator must also be on the label.

NFPA warning signs   The NFPA is an international organization whose original goal was to promote fire protection. The NFPA has prepared a 16-volume set of National Fire Codes. One of the codes, NFPA 704M, is the code for showing hazards of materials. It uses a diamond-shaped label or placard with appropriate numbers or symbols relating to fire, health, reactivity and special hazards (see Figure 9-3). Scales of 0 to 4 (low to high) classify the degree of hazard. This system also classifies the type of fire hazard and the type of extinguisher necessary to combat the particular fire type. Some counties require the NFPA signs, while others do not.

FIRE HAZARD (RED)

0 - WILL NOT BURN
1 - WILL IGNITE IF PREHEATED
2 - WILL IGNITE IF MODERATELY HEATED
3 - WILL IGNITE AT MOST AMBIENT CONDITIONS
4 - BURNS READILY AT AMBIENT CONDITIONS

HEALTH HAZARD (BLUE)

0 - ORDINARY COMBUSTIBLE
    HAZARDS IN A FIRE
1 - SLIGHTLY HAZARDOUS
2 - HAZARDOUS
3 - EXTREME DANGER
4 - DEADLY

REACTIVITY (YELLOW)

0 - STABLE AND NOT REACTIVE
    WITH WATER
1 - UNSTABLE IF HEATED
2 - VIOLENT CHEMICAL CHANGE
3 - SHOCK AND HEAT MAY DETONATE
4 - MAY DETONATE

SPECIFIC HAZARD EXAMPLES

OXY - OXIDIZER
ACID - ACID
ALK - ALKALI
COR - CORROSIVE

Figure 9.3 National Fire Protection Association.

State warning signs and labels The California EPA hazardous waste signs, labels and placards pretty much follow EPA's, mentioned above. Cal EPA also administers Prop 65, which contains requirements for warning employees and the public of the presence of certain toxic materials. If an exposure to one of the materials published on the state's list results in an exposure above that allowed by the law, the warning is required. The warning can be done by sign, hazard communication program, newspaper and other methods.

OSHA warning signs Various signs are required by OSHA to protect employees. Examples include exit, no smoking, caution and fire hazard. The OSHA signs are the most commonly seen signs.

Emergency Capability

This topic was mentioned in Chapter 2 in reference to the HMMP, SPCCP, HMBP, RMPP and Hazardous Communication Plans. These plans help establish a capability in organizations to deal with emergencies when

HMW are stored and used. These plans deal with emergency response, contingency and emergency preparedness. A small-quantity generator (SQG) is usually not required to have a written contingency or emergency plan; however, a large-quantity generator (LQG) must have one. Fire departments and state agencies such as the California Office of Emergency Services are two types of agencies that commonly get involved. The organization preparing the plan must think through what could happen with the HMW on the site. The plan should be continually updated. Key components of these plans include absorbents, neutralizers, HMW inventory, facility storage maps, fire extinguishers, blankets, training, emergency names and numbers, eye wash and showers [Allegri 1986].

Physical Design of Storage Areas

Facilities must be designed to minimize the possibility of a fire, explosion or any unplanned sudden or non-sudden release of hazardous waste or material to air, soil or water. The following design considerations apply to both hazardous materials and wastes.

- Store ignitable and reactive wastes at least 50 feet from the facility property lines (40 CFR 264.175).

- HMW should be stored away from traffic, including heavy forklift and foot traffic.

- A sufficiently impermeable base should underlie the containers to hold any leaked or spilled wastes or accumulated precipitation.

- Adequate ventilation should be provided to prevent buildup of gases.

- Drums should not be stacked more than two high. Flammable liquid drums should not be stacked at all.

- Eye wash stations must be provided for each storage area.

- Local agencies, such as the fire department, should be contacted prior to the design and building of a storage structure to incorporate other specific requirements. These agencies will probably require permits.

■ Container storage must be designated and operated to contain leaks and spills. This may mean epoxy-coated secondary containment. Local regulatory agencies, such as the fire department, may specify certain containment requirements. If only one container is stored, then 150% secondary containment should be provided. For multiple container situations, provide 110% of the largest vessel or 10% of the total volume to be stored. For outdoor storage facilities, the maximum probable quantity of runoff is also to be considered.

Oxidizers	Water Reactive	Irritants
Compressed gasses (except Oxidizers and Acetylene)	Flammable and combustible liquids	Bases
Acids	Herbicides and pesticides	Oxidizing compressed gasses
Radioactive materials	Carcinogens and bioaccumulatives	Reducers and toxic chemicals
Explosives	Acetylene	Etiologic agents

Figure 9.4 Storage of Chemicals and Wastes in Separate Compatibility Areas.

■ Incompatible HMW should not be stored together. They should be kept separated by barriers such as walls or berms. Figure 9-4 shows one possible way of segregating common chemicals into major compatibility areas. Each area shown needs to be either a separate storage location or divided by walls, berms and at least four feet of distance. There are actually even more compatibility subdivisions possible if every possible reaction were to be avoided. The 15 presented are the major groups that should be considered at a minimum. It is desirable to position the most incompatible chemicals as far from each other as possible even if they are separated by walls, such as acids from bases and oxidizers from flammables.

■ A drainage system should be installed so that spilled wastes or precipitation does not remain in contact with the containers. Alternatively, a mechanism such as elevation could be installed to minimize contact between spilled waste liquids and the containers. By storing HW drums on elevated platforms or pallets, they can be inspected and moved with greater ease. Storm water runoff into the containment system must be prevented unless the collection system has sufficient excess capacity, including a 10% volume buffer. Spilled or leaked waste and accumulated precipitation must be removed from the sump or collection area in as timely a manner as is necessary (usually eight hours) to prevent overflow [Baert 1990]. The collected HW should then be properly managed.

One commercial HMW storage and transfer area the author observed in Malaysia in 1991 sums up the way HMW should *not* be stored. This facility received HMW from many different industries and stored them all in the same area. The drums were stored on soil in the open air. Due to rain, many were rusted and leaking. Tankers were also present, with considerable soil-staining around the vehicle. Within 100 yards of this commercial facility was a fish farm and village.

Duration of Storage

It is easy to say how long you can store a hazardous material, and a very different and cumbersome answer for hazardous waste. If it is a hazardous material that is being used or will be used in the future, it can be stored indefinitely if the product is still in usable condition.

Hazardous waste storage is different. In general, a generator may accumulate HW in their primary storage area for up to 90 days without a TSDF permit, if the containers are in compliance with storage facility requirements, such as those for proper labeling. The 90-day clock starts with the first addition of waste to the container.

In certain states, if less than 55 gallons of hazardous or one quart of extremely hazardous waste are accumulated during any month, the waste can be stored for up to one year. The 90-day clock starts when these amounts are exceeded or the year is up.

Very closely related is a special provision in California called Satellite Accumulation (22 CCR). The storage container must be at or near the point of generation. It must be under the control of the operator of the process that generates the waste. The 90-day time clock does not start until the waste is transferred to the main HW storage area.

Duration of storage of hazardous waste in other parts of the world is significantly different from the United States. For example, in some parts of Europe such as Portugal and in the Far East, hazardous waste can be stored indefinitely. This is a practical solution to the severe shortage of TSDFs. Obviously, the waste is to be stored in a safe manner.

Unforeseen, temporary and uncontrollable circumstances are treated as special cases. Extensions to the 90-day period are possible in these situations. Usually, 30 additional days would be granted on a case-by-case basis. Storage details are summarized in Table 9-2.

Table 9.2 Hazardous Waste Storage Permit
and EPA ID Number Requirements—Federal Law

■ *Less than 100 kg Hazardous Waste or 1 kg Acutely Hazardous Waste Generated Per Month (Conditionally exempt Small-Quantity Generator)*

- No permit or ID number required
- When total quantity of waste exceeds specified limits, the generator has three days to comply with standard Small-Quantity Generator requirements

40 CFR § 261.5

■ *Between 100 kg and 1000 kg of Hazardous Waste Per Month (Small-Quantity Generator)*

- EPA ID number required (40 CFR § 262.12)
- No permit required if

   (1) Store up to 180 days; do not exceed 6000 kg of hazardous waste on site at any one time; if comply with specific operational requirements (40 CFR § 262.34 (d)); or

   (2) If have to transport waste over 200 miles can store up to 270 days without a permit (40 CFR § 262.34 (e))

- Need permit if exceed time limits, maximum storage or fail to meet operational requirements

■ *Over 1000 kg of Hazardous Waste Per Month (Large-Quantity Generator)*

- Need EPA ID number (40 CFR § 262.12)
- Can store up to 90 days without a permit if meet certain operational requirements
- If store over 90 days or do not meet operational requirements need a permit

40 CFR § 262.34 (a)

Table 9.2 Hazardous Waste Storage Permit
and EPA ID Number Requirements—Federal Law (Continued)

■ *Satellite Accumulation at Point of Generation*

- Rules on permits and ID number are the same as set forth above

- Up to 55 gallons of hazardous waste and one quart of acutely hazardous waste can be stored at the point of generation if certain operational requirements are met

- Once these limits are met, the generator has three days to move the waste to its standard storage area

- Standard time limitations start to run as soon as the limit is reached, not when the waste is moved to the standard storage area

40 CFR § 262.34 (c)

Storage Container Characteristics

Since most HW will be shipped to a TSDF, generators usually store HW in DOT-approved containers. The HMW storage container does not have to meet DOT's shipping container specifications until transportation on public roads occurs. It must, however, be sound, in good condition, free of leaks, stored in a containment area, kept closed during storage, inspected weekly and contain HMW compatible with the container. If a container is reused it must be triple-washed if it previously held an incompatible waste or material. The washing may require a permit and may also generate a hazardous waste.

The electronics manufacturing industry generates large volumes of low toxicity waste, such as soiled rags, polishing tape and nickel sludge. Therefore, the containers must be large and still covered and marked. This may include roll offs and 55-gallon drums, for example.

Inventory Management

Without thinking, most people negatively impact the environment and themselves by not practicing inventory management. When this is done, many people store and use a larger variety of products than needed. Inventory management techniques will help remedy many HMW problems. It is best to not order or store an excessive number of similar products from different manufacturers. For example, many organizations have similar hand cleaners and solvents from a multitude of different manufacturers. This usually results in duplicate orders, excessive volumes on-site and some chemicals never being used. Another good inventory management technique is to use a first-in-first-out inventory method. This involves the use of the older products, which have been sitting on the shelf, before using the newer items. This will prevent decomposition and disposal of unused HMW for exceeding shelf-life. If there are still too many chemicals being stored, they should be advertised as surplus stock on electronic mail systems or sent back to the main storeroom or manufacturer for reuse by others.

Case Study

Union Carbide—Bhopal, India

On December 3, 1984, one of the world's worst accidents occurred in Bhopal, India. 2000 to 3000 people were killed with over 200,000 others injured. The exact number of people involved, amount of chemical released and other details are a matter of continuing controversy.

The probable source of the problem was a leak of approximately 40 tons of a lethal and volatile chemical named methyl isocyanate (MIC). This chemical was being manufactured by a subsidiary of Union Carbide. MIC is a toxic ingredient used in the production of certain pesticides such as Temik and Sevin. Possible entry of water into an MIC tank led to uncontrolled build-up of heat and pressure. The chemical reaction that resulted could not be stopped because of nonfunctional safety systems.

The various entities involved, such as Union Carbide, the government of India, and the representatives of the dead and injured, are suing each other in court. Billions of dollars are at stake. Since the accident, Union Carbide has sold off many of its assets and is now a much smaller company than at the time of the tragedy.

This chemical calamity made industry and lawmakers around the world look at hazardous chemical storage and handling very carefully. It has resulted in the elimination of large stores of hazardous chemicals, and the creation of emergency planning and community right-to-know standards. It has also caused organizations to systematically identify hazards and use risk assessment techniques.

Union Carbide is presently using a very rigorous system of risk assessment methods. Hazards are identified and a risk review completed every four to seven years. The risk assessment team makes decisions to shut down, alter or leave a process in place [EPA 1989].

CHAPTER

TEN

Shipment of Hazardous Materials and Waste

Introduction

"One operation that allegedly runs out of Hartford, Connecticut only works in foul weather. A driver watches the forecast for rain or snow, then picks up a tanker load of chemicals. With the discharge valve open he drives on an interstate until 6,800 gallons of hot cargo have dribbled out. About 60 miles is all it takes to get rid of a load, boasted the driver, and the only way I could get caught is if the windshield wipers or the tires of the car behind me start to melt" [Block 1985].

Purposeful dumping and accidents routinely occur during the shipment of hazardous materials and wastes. Oil tankers carrying raw crude oil have spilled millions of gallons into the world's oceans. These accidental spills violate transportation and water-quality regulations, such as the Clean Water Act. The Clean Water Act gives the U.S. Coast Guard and the EPA authority when accidents and spills impact U.S. and coastal waters. The DOT is the primary agency involved in the regulation of hazardous material transportation.

This chapter will present information concerning shipment of both hazardous materials and wastes, including up-front activities, packaging materials and containers, labels and placards, markings, paper work, care in loading and variances. This chapter will stress shipment by truck since most HW is transported in that way.

On December 21, 1990, the DOT published the final rule for Docket HM-181, often referred to as HM-181 or the Performance Oriented Packaging (POP) regulation. This has changed major components of the way DOT regulates the transportation of hazardous materials. Compliance with the new regulations started on October 1, 1991, and will be complete on October 1, 1996. There are certain requirements that need to be met on October 1 of 1993, 1994 and 1996. Documented training of HM workers was required by October 1, 1993 and old DOT specifications for containers will be prohibited starting October 1, 1996. The most significant change is the conversion to the international hazard classification system. Another important change requires that the shipper determine and assign a packaging group designation as follows: I—great degree of danger, II—medium degree of danger, III—minor degree of danger.

Regulation

Shipment is primarily regulated by the DOT under HMTA, and the EPA under RCRA, and usually occurs via highway, rail, water and, in some cases, by air. HW is usually transported only by highway. The contractor must be licensed, registered/certified and have the proper permits. EPA and DOT regulations concerning transportation of hazardous materials are similar. EPA adopted DOT regulations in 1980 for labeling, marking, placarding, containers and discharge reporting. Also in 1980, DOT amended its hazardous materials regulations to make them applicable to hazardous waste. EPA and DOT regulations are found in 40 and 49 CFR, respectively.

Great care must be exercised when it comes to shipping a hazardous material or waste. This care applies to the initial shipment of chemical, shipments by customers returning some unused chemicals, shipments by customers from one site to another and shipment of waste. In other words, all shipments of chemicals and wastes are regulated above threshold quantities. EPA, U.S. Coast Guard, DOT and other agency regulations must be followed to the letter, especially packing, labeling and paperwork. If these are not correct, large fines can result and will increase significantly if a problem develops in transit, such as a spill.

Up-Front Activities

Many initial transportation actions should occur early, months before the shipment date. Most of these activities have already been discussed.

For example, a determination of whether the material or waste is hazardous according to the EPA (40CFR 261 and 262.11), DOT (49 CFR 172.101) or state must be made very early in the entire process.

Second, if a hazardous waste is involved, the destination must grant approval of the shipment prior to transportation from the generation site. In addition, an EPA ID number should have been obtained and shown on the manifest. The generator should also verify that the transporter and TSDF have EPA ID numbers (40CFR 262.12). A determination of whether any additional or special shipping requirements apply should also be made early by checking other references, such as 49 CFR 174-177.

Packaging Materials and Containers

Packaging, marking, labeling and container requirements are found in a variety of references. EPA marking requirements are in 40 CFR 262 and 263. DOT requirements can be found in 49 CFR 172, 173 and 179. There is some flexibility in terms of packaging materials, but not in labels and containers.

A wide variety of packaging materials are available for shipping HMW, most of which have two or three basic characteristics. The material must be inert and not react with the HMW. Second, the material must be absorbent or able to consume moisture and liquids. Last, the material must be able to absorb impact in case the container is bumped or dropped. Vermiculite is one example of a common packaging material. Packaging may be reused, if it is not contaminated during the transportation process. If a liquid is packed in a DOT-approved box, there must be enough absorbent present to contain the total volume.

Hazardous materials and wastes can be shipped only in DOT-approved containers as specified in 49 CFR 178–179. There are a few exemptions, such as for limited quantities. Most hazardous material container specifications and regulations are issued by the Research and Special Programs Administration of the DOT. These regulations apply to drums, dumpsters, in bulk by vessels, tank cars, portable tanks, cylinders, barrels, cans, boxes, bottles, casks, bags, tank trucks and rolloffs.

The design and operation standards for containers are a direct response to the historically haphazard usage of containers for storage and shipment of hazardous wastes. An ideal storage and shipment container must be in good condition. This includes being free of leaks, structural defects or rust. The outside should be clean and without holes, dents, bulges or cracks. The container should remain closed at all times, except when addition or removal is to take place [Baert 1990].

The most common container for HW is the 55-gallon drum. Drums going to Class I landfills must be 90% full and have no visible liquids. They must have a sealing gasket and ring that will not be deteriorated by the contents.

Markings on the bottom of the drum should be checked to make sure the drum meets the regulatory requirements (49 CFR 172.101). With the old marking system, which will be phased out by 1996, a drum may show: 19 55 92 DOT 17E NATCO 9 STC. This indicates that the gauge of metal is 19; the drum's capacity is 55 gallons; it was manufactured in 1992; it has a DOT specification of 17E; the drum was manufactured by NATCO in the ninth month; and it is a single-trip container. Of all that information, the 17E is the most important.

Labels and Placards

The generator is responsible for supplying the correct labels and placards and providing these to the transporter. In many cases, however, the reverse usually occurs. Labels and placards are placed on shipping containers and transportation vehicles to indicate what is in the load in case of an accident or spill. They also help the handlers in making determinations of the degree of caution to exercise in their daily job. Figure 10-1 is an example of how information is presented on labels and placards. The symbol is placed at the top. In the center is the hazard class name or four-digit UN specific material number. At the bottom is the UN class number.

Labels must be four inches square and placed on individual containers of HMW that are less than 640 cubic feet in volume. The words or symbols on the labels are strictly regulated. See Appendix B for examples.

The labels are normally used for transport of hazardous materials in the United States. CFR 49, parts 100–177, present the regulatory requirements. The labels must be placed on the surface of the package near the proper shipping name. If more than one label is required, the labels should be placed side by side.

Placards are 10-3/4 inches square and are to be placed on HMW containers with a volume greater than 640 cubic feet. 49 CFR, subpart F, section 172.512, delineates special requirements for containers with a volume greater than 640 cubic feet. Examples of placards are shown on in Appendix B. The same "Poison and Inhalation" markings apply to placards as

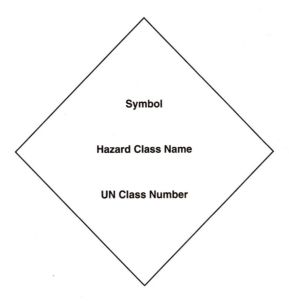

Inside the diamond, from top to bottom:

Symbol

Hazard Class Name

UN Class Number

Figure 10.1 Hazardous Material Labels and Placards. If hazardous materials are transported in tank cars, cargo tanks or portable tanks and if a specific UN/NA four-digit identification has been assigned, it must be used on placards in place of the hazard class name or printed on an orange panel.

they do to labels. If the gross weight of all hazardous materials is less than 1000 lbs., no placard is required for land transportation. The Dangerous When Wet materials are an exception to this, and that particular placard must be displayed for any quantity transported.

Various regulatory bodies cover international shipment. Shipments on water are controlled by the International Maritime Organization. Shipments by air are regulated by the International Civil Aviation Organization. In addition, most of the placards and labels are based on United Nations recommendations and many are wordless. With that exception, most are similar in color and symbols to the U.S. DOT requirements.

Only specified colors, numbers and symbols can appear on labels and placards. The color and class number system used on labels and placards is as follows:

■ Explosives—orange, #1

■ Compressed gases

  • Non-flammable—green, #2

- • Flammable—red, #2

- • Poison—white, #2

■ Flammable liquids—red, #3

■ Flammable solids—white with red stripes, #4

- • Dangerous When Wet—blue, #4

- • Spontaneous combust—white top, red bottom, #4

■ Oxidizers and organic peroxides—yellow, #5

■ Poisons—white, #6

- • Harmful—white, #6

■ Infectious substance—white

■ Radioactive materials—yellow and magenta, #7

■ Corrosive materials—white and black, #8

■ Miscellaneous - white and black, #9

■ Empty—black and white

Other Markings, Identifications and Numbers

The EPA requires that generators mark hazardous waste containers with ≤110 gallons with a statement that federal law prohibits improper disposal of hazardous waste. See Figure 10-2 for hazardous waste shipment marking. The label must also include proper shipping name and UN number. The UN number for hazardous wastes in general is 9189. This is used for chemicals shipped under the Hazardous Waste shipping name that don't fall under a specific hazard class.

To aid in identification of some hazardous materials, a specific product ID number is used in addition to the color and UN class number system. These can be found in the middle of the placard or on orange panels. The numbers are also UN-regulated and are four characters. For example, acetone has a specific product UN number of 1090. This specific product ID number must be used when transporting hazardous materials in tank cars, cargo tanks and portable tanks or, in other words, considered "bulk" by DOT.

```
◆◆◆◆◆◆◆◆◆◆◆◆◆◆◆◆◆◆◆◆◆◆◆◆◆
HAZARDOUS
WASTE
STATE AND FEDERAL LAW PROHIBIT IMPROPER DISPOSAL.
IF FOUND, CONTACT THE NEAREST POLICE OR PUBLIC SAFETY
AUTHORITY, THE U.S. ENVIRONMENTAL PROTECTION AGENCY
OR THE CALIFORNIA DEPARTMENT OF HEALTH SERVICES.
GENERATOR INFORMATION:

NAME _____

ADDRESS _____ PHONE _____

CITY _____ STATE _____ ZIP _____

EPA  /MANIFEST
ID NO./ DOCUMENT NO. _____ / _____

EPA                    CA              ACCUMULATION
WASTE NO. _____ WASTE NO. _____ START DATE _____

CONTENTS, COMPOSITION: _____

PHYSICAL STATE:  |HAZARDOUS PROPERTIES:  ☐ FLAMMABLE  ☐ TOXIC
☐ SOLID ☐ LIQUID |  ☐ CORROSIVE ☐ REACTIVITY ☐ OTHER _____
_____
_____
_____
     D.O.T. PROPER SHIPPING NAME AND UN OR NA NO. WITH PREFIX
HANDLE WITH CARE!
           STYLE WMCA8
◆◆◆◆◆◆◆◆◆◆◆◆◆◆◆◆◆◆◆◆◆◆◆◆◆
Printed by Labelmaster, An American Labelmark Co., Chicago, IL 60646  (800) 621-5808        REV. 5/92
```

Figure 10.2 Hazardous Waste Shipment Label. (Reprinted with permission by: LabelMaster, Div. of American LabelMark Co.)

Paper Work

No carrier may accept a shipment of HMW unless a properly prepared shipping paper is present. For hazardous material, it is a bill of lading; for hazardous waste, it is a uniform hazardous waste manifest (manifest). Both of these were first presented in Chapter 2. Shipping papers should be in the cab of the truck and within reach.

The person needing to transport the substance must certify accuracy of the papers and markings. Special certifications are required if the shipment is by air, especially radioactive materials [Allegri 1986].

The normal manifest procedure involves the preparation of a separate manifest for each generator, each site, and each shipment of hazardous waste. It is prepared by the generator and given to the registered hauler with the waste. The hauler signs and gives a copy back to the generator. A copy is sent to the regulatory agency at this point. The hauler and disposal site keep a copy. After the disposal site has accepted the waste, the disposal site sends a copy to the regulatory agency (allowing a match with the copy sent by the generator), and a copy goes back to the generator (also allowing a match). The generator has up to 45 days to notify the agency if the manifest copy has not arrived from the TSDF.

Under special situations, a modified manifest procedure may be approved. For example, an agency may approve the use of one manifest for the transport of waste oil from multiple sites of one generator. In this case, a record is kept of volumes from all sites, and at the end of a day the generator completes one manifest. Also, with special agency approval, the same concept may be possible with other wastes such as solvents and sulfuric acid.

Another special category involves a cover manifest. This is prepared by a TSDF. HW from various generators can be stored at a TSDF and then shipped together. When they are shipped for final disposal, the cover manifest is prepared by the TSDF and accompanies the individual generator's manifests.

In addition to preparing a manifest or bill of lading, the shipper has other record-keeping and reporting requirements. Many of these can be found in 49 CFR 172 and 40 CFR 262, Subpart D, and have been discussed earlier. Land disposal restriction (LDR) notification and certification, biennial reports, waste analysis and exception reports are examples of other forms to submit and record-keeping requirements related to shipment of HW.

The transporter must also maintain certain emergency response paperwork. This information may be attached to the manifest or be part of an MSDS or a separate emergency response manual. The DOT's Emergency Response Guidebook is usually referenced.

In Europe, the shipment of hazardous waste is covered by a Consent Order. It is similar to a manifest; however, there are fewer copies. If the waste is sent between different countries, additional paperwork is needed and considerable time. For example, it took the author close to two years to get permission to ship hazardous waste from Portugal to the United Kingdom. This was necessary because there were no disposal sites in Portugal at the time.

Care in Loading and Transportation

Since a very high percentage of HMW accidents are transportation-related, it is important to summarize a few loading and transportation considerations. One of the most important is proper training of handlers, load-

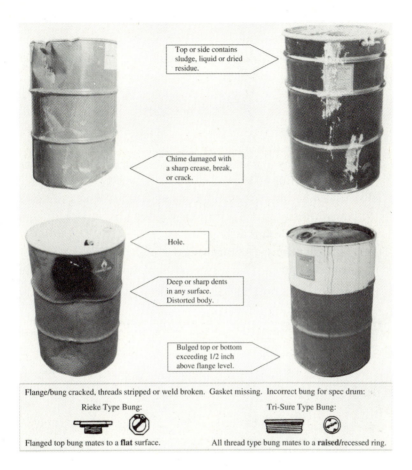

Figure 10.3 Container Loading and Transportation: Illustrations of Unacceptable Conditions. (Reprinted with permission from Romic Chemical Corp., 1994, East Palo Alto, California.)

ers and drivers. This is part of the HM-181 and HM-126F training requirements effective October 1, 1993. Three eight-hour classes are required and must cover topics including classification, shipping name, marking, labeling, placarding, documentation, limited quantities, ER, RQ, packaging and container conditions (Figure 10.3).

Many of the loading and unloading considerations are rather obvious. During loading and unloading, the brake must be set and the engine stopped. There should be no smoking. The personnel should inspect for sharp projections and problem flooring. All tailgates, closures and the cargo must be secured. Quantity limits are not to be exceeded. The person handling the shipment should refrain from drug or alcohol consumption.

Containers deserve special attention during loading and transportation. The containers must be handled so as to prevent rupture, leakage or spills. They should be routinely inspected for integrity on a frequent basis. If the inspections reveal a leak or deterioration of the container, the HMW must be transferred to another container [Baert 1990].

When transportation is occurring, even more safety considerations come into play. For example, the driver cannot exceed 10 hours/day or a maximum of 60 hours in each seven consecutive days. The cargo must be attended by someone at all times and its temperature maintained within a prescribed range. At set frequencies, the cargo must be reinspected. The cargo should be kept dry.

After haulage, decontamination of the vehicle may be required. This especially applies if the cargo contained radioactive materials or asbestos [Allegri 1986].

Transportation Exceptions and Variances

A commercial transporter can temporarily store a loaded shipment of hazardous waste at a transfer station for up to 10 days. If additional time is needed, the waste must be taken to a permitted TSDF.

A generator can transport their own waste on-site without special permits. This must be done carefully, however, and cannot involve a public road that may be present on-site or between sites.

Under special situations, which must be approved by the agencies, certain transportation variances are possible. These variances may remove some of the requirements, such as manifesting. For example, small volume, short distance or low toxicity may reduce certain shipping requirements.

Case Studies

France

France bans 700,000 tons/day of German waste containing hazards. France has charged three company officials with illegal importation of medical waste. France has also banned the movement of all trash from Germany. Trucks carrying hospital waste disguised as household refuse were stopped near the German border. France feels it has become a dumping ground for Germany, whose rigid environmental laws make disposal of waste in Germany cost-prohibitive. The ban on German trash by France was published on August 18, 1992, and took effect the next day [Adams 1992].

Sumter, South Carolina

All packed up and no place to go! After being loaded with 2400 tons of contaminated soil from an earlier spill, a train roamed the tracks for over a month searching for a disposal site for its hazardous waste. During its aimless wanderings, the train was said to be leaking some hazardous liquids. Part of the reason the train continued to roam was because an environmental group was reported to be scaring off potential landfill operators from accepting the waste. After plowing through eight states, the train was last seen rolling out of Sumter, South Carolina, to an undisclosed destination [Time 1991].

CHAPTER

ELEVEN

Treatment of Hazardous Wastes

Introduction

Considering the volume of hazardous waste generated, there is a serious shortage of treatment facilities in many parts of the world. This is in part due to resistance to the siting and operation of these facilities or "not in my back yard" (NIMBY) syndrome, and to complex regulatory requirements. As of 1991 there were 4493 commercial treatment facilities (including disposal) permitted under RCRA in the U.S. A partial breakdown of this number is shown in Figure 11-1. The largest number of facilities are treatment and storage. This number would therefore be made up of facilities that do concentration-type treatment and detoxification treatment (except incineration, which is a separate number).

Regulations, permits and standards that apply to TSDFs have been partially covered in Chapter 2. In general, the TSDF must comply with local, state and federal standards that concentrate on the monitoring of ground water, financial accountability and closure/post-closure processes. For example, four ground water monitoring wells are normally required under RCRA for the life of the facility plus 30 years. Financial assurance of a facility's ability to implement closure and post-closure must be presented. The closure process for the facility must include decontamination of structures, placement of final cover and other activities.

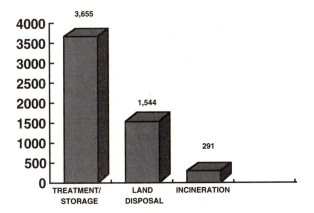

Figure 11.1 Number of Active Facilities by Type of Treatment in 1991. (From EPA, Hazardous Waste Data Management System, Washington, 1991.)

When making a treatment technology determination, several general considerations are important. At the top of the list are the technical/efficiency, cost, worker safety, regulatory and public acceptance considerations. The size of the treatment unit is also important and can be of the small/single-user type all the way up to large/multi-user units. Treatment systems can be transportable or of the permanent in-place variety.

The design of treatment systems can become extremely complex, especially if different types of waste are to be handled. Conceptual, preliminary and detailed design steps are usually carried out for most systems of any size. If the system is new and unproven, it may be first tested in the laboratory (bench or jar tests), or a working model or pilot plant (prototype) may be constructed before the actual system is built.

The physical location of the treatment facility can be either off-site or on-site. If treatment occurs off-site, it is commonly at a sewage treatment facility or a specialized treatment facility and is for multiple generators. If treatment occurs on-site, it is usually by a single generator. Transportable treatment units (TTUs) can be brought in for certain types of treatment and serve one user for a period of time and then are hauled to another site until the contamination is cleaned up there. Whenever possible, the treatment should be on-site, since it usually reduces the overall cost and potential liability [Wentz 1989].

Treatment technologies can be classified in many ways. The categories used in this chapter include concentration, detoxification and fixation technologies. Concentration achieves volume reduction and is generally less expensive. Detoxification destroys toxic components or reduces them to a less toxic form and is generally more expensive. Fixation technologies reduce the mobility of the toxin. Many waste streams use a combination of several types of treatment. For example, in recovering solvents and oils many of the treatment technologies are used. A few of the treatment technologies were previously presented in Chapter 8 during discussions of soil and ground water treatment.

Concentration Technologies

The primary goal for concentration technologies is to reduce the volume of the hazardous waste. This is generally done before detoxification. Concentration technologies are usually a physical form of treatment. Air stripping and oil/water separation are also concentration technologies, but have already been discussed in Chapter 8 concerning ground water treatment.

Concentration technologies can be generally categorized into three groups. These groups include gross removal of contaminants, solids dewatering and polishing. Gross removal of contaminants would be done by precipitation, sedimentation, flotation, clarification and oil-water separation technologies. An example of solids dewatering would be centrifugation. Polishing includes filtration, evaporation, ion exchange, peat adsorption and reverse osmosis. These categories are not hard and fast, and one technology may serve in all three groups.

Precipitation

Inorganic aqueous wastes are treated by precipitation; many dissolved heavy metal ions (iron, potassium, etc.) can be chemically precipitated and then physically removed. Lime, sodium hydroxide and sulfide salts are added to a solution to adjust the pH to a point that causes precipitation. The precipitation usually occurs in the form of hydroxides or sulfides. On the left side of Figure 11-2, the precipitation chemicals and flocculation aids are being added. In the second tank, the actual flocculation process is underway, or the contaminant is forming into larger units that will settle out in the final tank as a metal sludge.

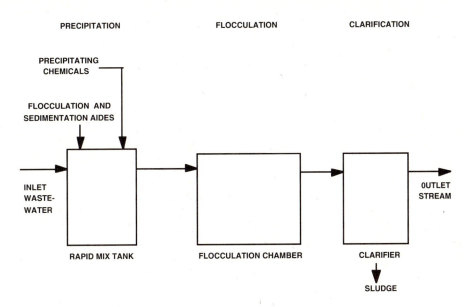

PRECIPITATION **FLOCCULATION** **CLARIFICATION**

Figure 11.2 Schematic Diagram of Chemical Precipitation and Associated Process Steps. (From EPA, *Mobile Treatment Technologies for Superfund Waste*, EPA Report 540/2-86/003(F), Washington, 1986.)

Sedimentation

Sedimentation, or gravitational settling, removes some particles from a liquid waste. It is a gravity-settling process that allows heavier solids to collect at the bottom of a containment vessel, resulting in their separation from the suspending fluid. Chemicals are added to cause flocculation or coagulation, which produces faster-settling particles. It may require up to 48 hours for settling to occur. The particles are carried out in sedimentation basins, clarifiers or inclined plate separators. Sedimentation is restricted to solids that are more dense than water, and it is not suitable for wastes consisting of emulsified oils.

Surface impoundments are considered a form of treatment by the EPA. Various processes occur in the traditional surface impoundment; however, sedimentation is the primary action (and, to a lesser degree evaporation). Even if the popularity of sedimentation ponds has waned with tighter regulations, they are still used to a certain degree after the proper permits are obtained. Regulatory requirements for ponds modified or constructed after November 8, 1994, include two or more liners, leachate collection between the liners, ground water monitoring, vadose zone monitoring and proper operation and maintenance. For example, the settled materials must be routinely cleaned out and treated or sent to a Class I disposal site.

Filtration

Filtration is used after precipitation and sedimentation to remove residual suspended solids that do not settle out. Filter medias can include silica sand, anthracite coal and garnet supported on beds of gravel, screen and various membranes. The filters are either single or mixed media. Depending on filter type, cleaning occurs via backwashing, or forcing water in the reverse direction through the filter. The three types of filtration systems used most often are granular bed, diatomaceous earth and membrane filtration. Filtration should not be used with sticky or gelatinous sludges due to the likelihood of filter media plugging.

Evaporation

Evaporation is used to concentrate and recover or remove salts, heavy metals, plating solutions and other solids and liquids. It is applied to solvent waste contaminated with nonvolatile impurities such as oil, grease, paint solids or polymeric resins. There can be reuse of the vapor (de-ionized condensate) and/or the heavy metals and salts left in the remaining solution. The types of evaporation include atmospheric and vacuum evaporation. In atmospheric evaporation, natural evaporation occurs or is accelerated by boiling the liquid, spraying a heated liquid on an evaporation surface or blowing air over the surface. In vacuum evaporation, the evaporation pressure is lowered to cause the liquid to boil at reduced temperatures (see Figure 11-3).

Ion Exchange

In this technology, ions are exchanged for ions of similar charge from the solution in which the resin is immersed. The process depends upon the electrochemical potential of the ion to be recovered vs. that of the exchange ion, and also upon the concentration of the ions in solution. The entire exchange process occurs on the surface of the resin. The process results in the recovery of rinse water and process chemicals. At the end of the process, regeneration of the resin, which is holding the impurities, occurs, or it is disposed of as a hazardous waste. Examples of industries that use ion

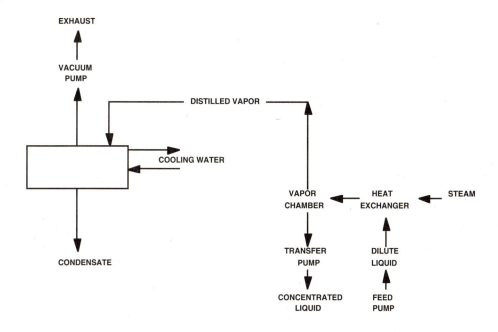

Figure 11.3 Schematic Diagram of Single-Effect Evaporators. (From EPA, *Mobile Treatment Technologies for Superfund Waste*, EPA Report 540/2-86/003(F), Washington, 1986.)

exchange include metal finishing, photography, chemical, food, nuclear, pharmaceutical and textile [Wentz 1989].

Peat Adsorption

Adsorption is used to treat many organic liquid hazardous wastes. One type of adsorption material is peat moss, which is used in the purification of waste waters from many industries. It affords good adsorption because of the polar nature of the material. Peat adsorption is used to remove dissolved solids such as metals and polar organic molecules to low levels.

Flotation

Flotation is the opposite of sedimentation. Suspended solids are caused to float to the surface of a tank, where they can be removed. This is commonly used to remove metal hydroxides and oil. It involves the release of gas bubbles or foam, which attaches to the particles, making them buoyant (see Figure 11-4). This technology is only applicable for waste having densities close to that of water.

Flotation systems can be used for finely mixed, suspended and non-soluble materials. Contaminated water is slowly passed through a tank that contains a coalescing grid. The materials stick to the grid and eventually form large units which float to the surface.

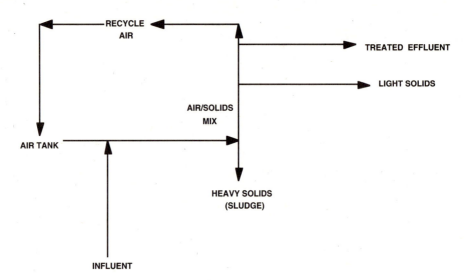

Figure 11.4 Schematic Diagram of Recycle Flow Dissolved Air Flotation System. (From EPA, *Mobile Treatment Technologies for Superfund Waste*, EPA Report 540/2-86/003(F), Washington, 1986.)

Reverse Osmosis

Reverse osmosis is a membrane system (as is dialysis) that can be used to concentrate ions on one side of a semipermeable membrane and solvent on the other. The ions or large molecules cannot pass through the membrane. High pressure is used to force the solvent to move to the side with the lower ionic concentration (the reverse of osmosis). This is used for various applications, such as in desalting sea water. For efficient reverse osmosis, the chemical and physical properties of the semipermeable membrane must be compatible with the waste stream.

Centrifugation

Solids can be concentrated utilizing the principles of centrifugal force. Differences in molecular weights allow for the separation of materials under centrifugal force. Treatment systems for both inorganic and organic liquid, sludge and slurry wastes utilize this type of technology. This is similar to the laboratory centrifuge, which concentrates cells in the bottom of a test tube, except on a larger scale. In industry this is often a continuous process, where the incoming stream is separated into two effluent streams, one of which contains most of the solid particles.

Solvent Extraction

Solvent extraction is used for the treatment of organic liquid, sludge and slurry wastes. The system helps in waste volume reduction since the organics are separated from water or nonhazardous solids. Clean solvent is recovered and reused in the system to extract the contaminants. This in turn reduces transportation and disposal costs. Solvent extraction is used in site remediation and fixed-facility systems.

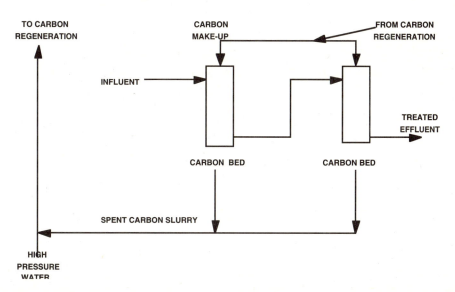

Figure 11.5 Schematic of Granular Activated Carbon Columns. (From EPA, *Mobile Treatment Technologies for Superfund Waste*, EPA Report 540/2-86/003(F), Washington, 1986.)

Activated Carbon

Activated carbon (see Figure 11-5) removes many organic and inorganic substances from liquids via adsorption on carbon atoms (a surface reaction). Carbon used for adsorption is usually treated to produce a product with large surface-to-volume ratio. The contaminant is incorporated onto a carbon matrix. The activated carbon is either in the granular or powdered form. The amount of surface area of the activated carbon determines the effectiveness in removing the hazardous constituents [Wentz 1989]. In Figure 11-5 the contaminated water is sprayed in the top of the first carbon bed. It then percolates through the carbon bed, allowing the contaminant to be adsorbed onto the carbon matrix. The process is repeated through the second carbon unit. Spent carbon can be regenerated, but for strongly adsorbed contaminants, the cost of such regeneration can be higher than simple replacement with new carbon absorber.

Detoxification Technologies

After the preceding systems have operated, there is usually an end product that is toxic and is usually treated by one of the systems mentioned in this section. The purpose of detoxification is to reduce toxicity. Most of the detoxification technologies involve oxidation in one form or another. Detoxification technologies are generally chemical or biological processes, rather than physical, as with concentration technologies. Common methods include biodegradation, chemical oxidation/reduction/neutralization and incineration. Most of these methods will not reduce the toxicity all the way to zero. Many variables impact which treatment method will work best. This fact is especially important in terms of detoxification technologies. For example, solubility, valence state and many other considerations determine which systems are employed.

Biodegradation, Biological Treatment and Land Treatment

Certain microorganisms (such as bacteria) can use diesel, PCB, oil and other substances as a food source. This was discussed in Chapter 8. Biodegradation can be accomplished on-site or hauled off-site to land farms where waste is incorporated into surface soil, degraded, transformed or immobilized through proper management. Of all the detoxification technologies, biodegradation is becoming one of the most popular. A great amount of research is being carried out concerning this technology, especially to genetically engineer bacteria that will more aggressively degrade a wider range of contaminants.

Biological treatment, similar to that done at a sewage treatment plant, can be used for certain hazardous wastes, such as organic contaminants. The process requires very controlled conditions; however, the microorganisms are usually sensitive to rapid increases in contaminant levels. These systems normally use one or more of the following subsystems: oxidation ponds, activated sludge, anaerobic digesters, aerated lagoons and trickling filters [Blackman 1993].

Land treatment, or land farming, is still being used to treat some hydrocarbon wastes and sewage sludge. Overall, however, this method is not as popular as it used to be because of regulatory restrictions such as treatment demonstrations, monitoring and other requirements.

Chemical Reduction, Oxidation and Neutralization

Chemical reduction involves the addition of strong reducing agents, such as sulfur dioxide and sodium bisulfite. Reduction is a reaction where the valence is decreased because of the addition of electrons. Hexavalent chromium waste is often treated in this way by causing the reduction to a trivalent form which will precipitate out. In Figure 11-6, the reducing agent is added to the chamber on the left, along with the waste and any pH adjustment chemical. In the last tank, a lime slurry is added to complete the reaction. Iron electrodes can also be used in place of the reagents to cause an electrochemical chromium reduction.

Chemical oxidation is used in the treatment of many inorganic and organic liquid and solid wastes such as cyanide. Oxidation involves a reaction where the valence is increased because of the loss of electrons. Oxidation of cyanide occurs by using chlorine, ozone, hydrogen peroxide or electrochemical oxidation. In addition to cyanide, phenols and other photographic industry wastes are also treated via chemical oxidation.

Chemical neutralization involves the simple neutralization of an acid by adding a base or the reverse. This is done both in treatment facilities and at spill sites (Figure 11-7). Neutralization is often used for both inorganic and organic liquid, sludge and slurry wastes. In Figure 11-7, the unit is set up to neutralize either acidic or basic waste. The waste being neutralized is piped into the waste storage area and then pumped into the neutralization chamber. If excess alkaline waste is not available for neutralizing acids, a lime slurry

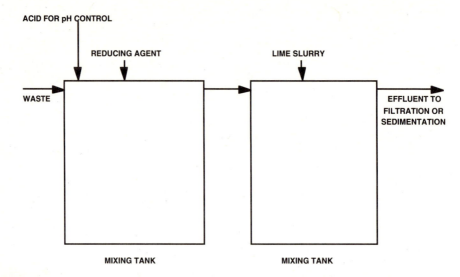

Figure 11.6 Schematic Diagram of Chemical Reduction. (From EPA, *Mobile Treatment Technologies for Superfund Waste*, EPA Report 540/2-86/003(F), Washington, 1986.)

or NaOH is added. On the other hand, if excess acid waste is not available to neutralize alkaline waste, fresh hydrochloric or sulfuric acid is added to complete the neutralization. Neutralization can be a very inexpensive treatment, especially if waste alkali can be used to treat waste acid.

Incineration

Incineration is also a detoxification treatment method that is used for many organic liquid, sludge and solid wastes. As with the other methods, incineration will not detoxify all materials. For example, metals that go into the incineration process will exit as hazardous substances in the ash. Infectious waste, on the other hand, is essentially detoxified during incineration (see Figure 11-8).

When considering incineration, attention to the primary hazardous waste and also the products of incomplete combustion (PICs) should occur. The latter are often not in compliance, especially with older incinerators. In these cases the three Ts of the incinerator environment may have to be adjusted: time, temperature and turbulence.

There are several types of hazardous wastes treated by incineration. These include organic, acidic, oily, halogenated compounds, drilling mud,

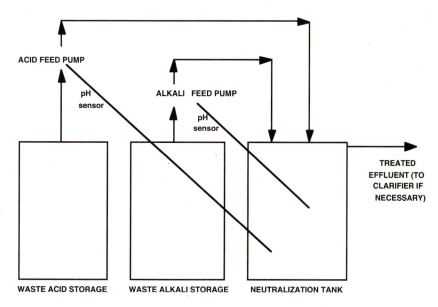

Figure 11.7 Schematic Diagram of Neutralization. (From EPA, *Mobile Treatment Technologies for Superfund Waste*, EPA Report 540/2-86/003(F), Washington, 1986.)

pesticides, PCB and pathological materials. Low-level nuclear waste (especially slightly contaminated trash) is also incinerated. The specific type of waste is one important factor in terms of which incineration technology should be used. Some incineration possibilities include liquid injection, rotary kiln, fixed hearth, circulating bed, multiple hearth, fluidized bed, infrared incineration, molten salt destruction, plasma torch and pyrolytic.

The boiler referenced in the middle of Figure 11-8 has outputs to an ejector system, the brine concentrator and the deaerator. A brine concentrator follows the neutralizer, which is shown in the lower right.

Air pollution is a concern related to incineration; air controls have been designed to minimize the atmospheric discharges. The primary air controls for incinerators are for either particulate or gaseous pollutants. Particulate controls include cyclones, baffles, fabric filters, electrostatic precipitators and scrubbers. Gaseous controls include afterburners, activated carbon and scrubbers.

A 45,000-ton/year RCRA hazardous waste incinerator is under construction in Kimball, Nebraska, by Waste Tech Services. As of August 1993, it was approximately 50% complete and expected to be finished by 1994. It will be the only commercial fluidized bed incinerator in the United

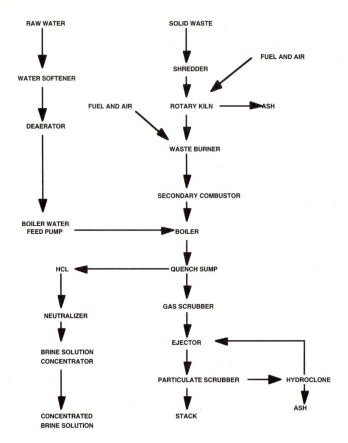

Figure 11.8 Simplified Process Flow Diagram of a Mobile Rotary Kiln Incineration System. (From EPA, *Mobile Treatment Technologies for Superfund Waste*, EPA Report 540/2-86/003(F), Washington, 1986.)

States. The ash generated will be disposed of at an on-site landfill as a non-RCRA hazardous waste.

Fixation/Solidification Technologies

Fixation or solidification reduces the mobility of inorganic hazardous chemicals. It involves fixing the toxins into a nonleachable matrix after the waste has been precipitated. For example, certain toxins can be mixed with cement, polymers, lime/fly ash or asphalt and are essentially encapsulated. The final composition of the mixture ranges from 50-60% fly ash to 10-

60% water to 5-15% lime to 8-20% waste. After the components are mixed, it is spread and allowed to solidify in drying beds [Blackman 1993].

Other wastes can be mixed with specialized substances and then subjected to a high current. The final product resembles glass. This melting or fusing process is called vitrification. The heat generated by the electric current causes melting that gradually works downward through the contaminated soil. Some contaminant organics are volatized and escape from the soil, and must be collected by a vacuum system.

Summary

There are hundreds of treatment technologies. Industry practice, however, has basically concentrated on four systems. Many organic wastes are treated by incineration or solvent extraction. Most metal-bearing wastes are treated by precipitation. Activated carbon is used for a variety of wastes in both the liquid and gaseous phases.

Case Study

Municipal Incinerator Ash

A federal appeals court's ruling that required cities to treat the ash from their municipal incinerators as hazardous waste rather than as ordinary waste has been overturned by the Supreme Court. This decision was based on the EPA's new policy, which exempts municipal ash from hazardous waste regulations. The average cost for disposing a ton of hazardous waste is $210, while nonhazardous material disposal cost is approximately $23/ton. This long dispute is based upon how Congress had intended the ash to be treated when it passed RCRA. Chief Judge William J. Bauer of the appellate court stated, "What we have to work with here is a statute subject to various interpretations, a foggy legislative history and a waffling administrative agency. Where do we turn?" [*San Francisco Chronicle* 1992].

C H A P T E R

TWELVE

Disposal

Introduction

The amount of HW disposal has been decreasing because of regulatory restrictions, costs and sincere interest in the environment. Public pressure has also had a hand in decreasing the amount of hazardous waste disposal. Incinerators, Class I sites and other treatment and disposal options have been severely resisted by environmental groups and other members of the community. The NIMBY syndrome will surely continue into the future and cause disposal to decrease while waste minimization increases. No matter how much treatment and waste minimization is practiced, there will always be some hazardous waste that requires disposal.

Responsibility for the waste at time of disposal is different for various countries. In the U.S. a generator retains a great degree of the responsibility and liability even after the disposal site has accepted the waste. In the United Kingdom the government holds the treatment and/or disposal facility responsible for the waste after they have accepted it.

It is pretty safe to say that there is a shortage of permitted hazardous waste disposal sites, no matter what part of the world you look at. As you can expect from earlier discussions, it is very hard to find an acceptable disposal site in the Far East. Thus, international shipment or on-site waste storage is commonly required. The situation is not quite as bad in Europe; however, there are countries, such as Portugal, without adequate sites. The U.S. at least has a few disposal sites, all of which are very expensive to use.

After the waste has gone through the treatment phase, it is ready for disposal. The form and type of the waste in large part determines where disposal will be allowed. A few limited liquid wastes can be disposed of by deep well injection and in the oceans. Solid-residue wastes are commonly landfilled. Many gaseous and particulate wastes are disposed of in the atmosphere.

In Europe, hazardous waste disposal has had a different evolution. Due to a general shortage of space, there have been few hazardous waste disposal sites established. In place of these sites, incineration has developed much more than in the United States. Also, large-scale recycling programs and waste-to-energy projects are, in general, further along than in the United States [Blackman 1993].

Land Disposal

Land disposal is regulated at the federal level primarily by the Federal EPA, and at the state level by agencies such as the California EPA and the California State Water Resource Control Board. The regulations concerning land disposal have changed dramatically over the last few years. In fact, recent regulations have promoted a shift away from land disposal. In October 1993, the EPA implemented additional requirements for landfills. This included new standards for closure, post-closure, bonding and financial assurance.

Many operators, such as in the electronics manufacturing industry, choose landfill waste in many cases instead of off-site recycling. This is due to the fact that sometimes it is quicker, cheaper and involves less liability than dealing with a recycler who may not have all the necessary permits and controls. Also, the landfill probably has many more customers (potential PRPs) to divide up a future cleanup. Unfortunately, this definitely is counter to preserving natural resources and saving valuable land fill space.

Classes of Land Disposal Sites

Class I—Waste Management Units for Hazardous Waste

These sites have become residual repositories and accept only hazardous waste treatment residues. Examples of full-service Class I sites in the Western U.S. include U.S. Pollution Control (Utah), Environmental Safety Services (Idaho) and U.S. Ecology (Nevada and Washington). Class I sites must be designed and constructed very carefully. Figure 12-1 illustrates some of the special care needed to prevent environmental contamination, such as the ground water monitoring wells, leak detection system, covers, liners and leachate collection systems. In addition to these, one would expect to implement air, wildlife, surface runoff, abandonment and other environmental and community protection systems.

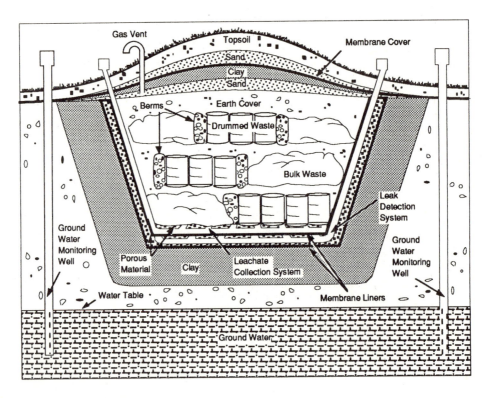

Figure 12.1 Schematic of Representative Hazardous Waste Landfill (not to scale). (Source: EPA, 1993)

Class II—Waste Management Units for Designated Waste These sites allow only certain solid hazardous wastes, such as soil with low-level contamination and asbestos, for which they are specifically designed. The Class II sites have some, but not all, of the extra control systems present in a Class I site.

Class III—Landfills for Nonhazardous Waste   These sites accept common household trash and construction debris and are normally called municipal landfills. In many cases, however, hazardous waste ends up in these sites because of error, ignorance, criminal intent or lack of concern. Whatever the reason, hazardous waste disposal in a Class III landfill is against the law and causes significant damage to the environment since the waste is simply buried with only a few control systems in place.

Land Bans

Land bans have been imposed at the state and federal levels for most untreated wastes. Land bans are actually more treatment standards than true bans. They require that wastes be treated to certain standards before land disposal can occur. These standards are either performance treatment standards (treat to x level) or technology driven standards. If the waste can't meet the standards for land disposal, it must be recycled, reused or treated further.

The primary legislation that resulted in land bans are the federal HSWA of 1984 and state legislation like California SB 1500 and the Hazardous Waste Management Act of 1986 (Health and Safety Code 25179.9-25179.11) The land bans apply to Class I land disposal sites, impoundments, waste pits and injection wells. Land bans apply to all RCRA and non-RCRA hazardous waste.

State land disposal restrictions   At the state level (e.g., California), land ban/treatment standard regulations can be found in Chapter 18 of 22 CCR. As of July 1989, all California listed liquid wastes were banned from land disposal.

California land disposal restrictions really got going with the ban on disposal of cyanide wastes greater than 1000 mg/l. Numerous other milestone dates have occurred since. A more recent important date specifies that as of May 8, 1990, generators were required to meet treatment standards for non-RCRA hazardous wastes. Very few extensions were granted, and now all wastes must meet the standards or are banned from disposal. The standards must be met prior to land disposal, disposal in surface impoundments or disposal into underground injection wells. Treatment standards apply to all generators, including small- and large-quantity. State standards are based on BDAT, STLC and TCLP.

Federal land disposal restrictions As with the states, the EPA has also banned untreated hazardous wastes from land disposal. In order to use land disposal, the waste must first be treated to reduce volume and/or toxicity and meet treatment standards. Federal land ban regulations can be found in 40 CFR 268.

The federal program has a couple of variance possibilities. In certain cases, a generator can get a variance upon petition to EPA if the TSDF can't treat or dispose of the waste. Sometimes a TSDF can also request a no migration variance and still accept wastes that don't meet standards.

The following 450 or more untreated hazardous wastes have been banned by the EPA (not a complete list):

May 8, 1995	Bulk liquids in landfills
November 8, 1986	27 organic solvent wastes
August 8, 1988	First third of all RCRA listed wastes— Applies to F006-F009, F019, 45 K wastes, 40 P wastes, 67 U wastes
November 8, 1988	Dioxin-containing wastes
June 8, 1989	Second third of all RCRA listed wastes— Applies to F010-F012, F024, 18 K wastes, 29 P wastes, 85 U wastes
August 8, 1990	Last third of RCRA wastes—Applies to 13 K, 38 P, 77 U, all D wastes, hard hammer for all wastes
November 9, 1992	Newly listed wastes
August 9, 1993	Ignitable and corrosive characteristic wastes whose treatment standards were vacated.
November 8, 1994	Additional wastes listed after this date will be evaluated on a case-by-case basis. Within six months of the listing, the EPA must decide whether land disposal of the waste is possible.

Ocean Disposal

Ocean disposal is primarily regulated under CFR Title 40, part 228. Interim dump sites are listed in Sec 228.12 of 40 CFR. The EPA and the Corporation of Engineers can issue permits for the dredging and dumping of certain wastes. There are many prohibited materials. In fact, most hazardous wastes have been prohibited by the Marine Protection Research and Sanctuaries Act. Some sludge and waste-water disposal at sea does still continue. For example, some off-shore oil platforms discharge production water, containing some pollutants, into the ocean.

RCRA does not apply to generators of HW at sea. For example, many ships discharge bilge water into the ocean. When ships anchor at California ports, however, they have to comply with California regulations, since bilge water is a HW is California.

To demonstrate acceptability for ocean dumping, the organization must perform special studies. These studies must show no significant adverse impact on marine life, chronic toxicity or bioaccumulation in marine life, after allowance for initial mixing. Many other conditions must also be met for the few wastes that can be disposed of in the oceans.

Deep Well Injection

Deep well injection is regulated primarily under the Safe Drinking Water Act and RCRA. It involves the pumping of liquid waste into underground porous formations, such as sand. These formations must be isolated from potable water. In the early 1980s, deep well injection was used very heavily in the U.S. and accounted for 8.60 billion gallons in 1981 [Wentz 1989]. That volume has decreased significantly, and presently deep well injection is used for limited disposal of some liquid hazardous wastes that can't be managed in other, more cost-effective ways. The waste can't be injected into or above potable water supplies. Considerable study is required prior to injection. During and after injection, monitoring must occur to ensure proper containment of the injected waste. The amount of deep well injection has been severely restricted due to a recent requirement that injected wastes meet the treatment standards of the land disposal restrictions.

The EPA established five classes of injection wells, labelled I–V. Class I has been used by industrial and municipal hazardous waste generators. A class I well must inject fluids below the lowest known drinking water source within 1/4 mile. Class II wells have been used to dispose of oil and gas well production fluids. Class III wells are used for the extraction of minerals. Class IV wells have been used for the disposal of hazardous and radioactive waste above a drinking water source and have been pretty much phased out. Class V includes all other types of injection wells [Wentz 1989].

The Toxic Injection Well Control Act of 1987 restricts use of deep well injection. This act prohibits injection of hazardous waste into or in communication with drinking water.

Atmospheric and Outer Space Disposal

Gases and particulates are disposed of into the atmosphere after treatment processes such as incineration and air stripping. The height of the discharge above the ground effects the amount of atmospheric mixing and dilution. Very tall emission stacks, such as at power plants, offer maximal mixing and dispersion. Wind conditions and ambient temperature also influence the effectiveness of the dispersion. At the other end of the spectrum, soil ventilation (to remove volatile organics) usually offers only minimal atmospheric dispersion.

Prior to the atmospheric discharge, many types of air pollution control equipment help to bring the toxic components into acceptable ranges. For example, cyclones, baffles, fabric filters and electrostatic precipitators help control particulates. Afterburners, thermal oxidizers, activated carbon and scrubbers reduce gaseous emissions.

In California, atmospheric discharge is regulated (via permits and emission credit systems) primarily by the Air Quality Management Districts for point sources. Non-point sources, such as automobiles, are regulated by the Air Resource Board. There is considerable competition for the permits and emission credits since they may mean the continued survival of an organization whose operation involves air emissions.

Outer space disposal of hazardous waste has been suggested by some. Most proposals include sending certain hard-to-dispose-of wastes, such as high level radioactive waste, toward the sun to allow incineration. This may have actually occurred on a small scale in the past, when space vehicles left their planned orbit and were pulled toward the sun. Other vehicles have been lost in deep outer space, along with their on-board hazardous materials and wastes.

Outer space hazardous waste disposal would need to be studied very carefully before being allowed to occur. Environmental impacts at the solar system level are undefined.

Case Study

Love Canal—Niagara Falls, New York

Niagara Falls, New York, was the location of great industrial expansion during the late 1800s. In 1892, William T. Love developed a plan to build a canal which would join the lower and upper levels of the Niagara River. Love intended to have a city developed along this canal to use the generated electricity. Construction was halted when new discoveries led to the transmission of alternating current. The canal was left with two sections unfinished.

Since power became more available, industries began to move into the area. In the 1930s Hooker Chemical and Plastic Corporation began landfilling at one section of Love Canal. By 1947, other companies were using the site for disposal. The site was filled with pesticides, plasticizers and caustic materials. At that time there were few disposal regulations, and the area was away from the population. Other materials, such as biological warfare agents and fly ash, were also deposited there until it was closed and capped by the owner, Hooker Chemical, in 1952.

In spite of warnings and a filed disclaimer by Hooker Chemical, the City of Niagara Falls demanded to purchase 16 acres of the land for $1.00. The City then built an elementary school over the closed landfill in 1953. The remainder of the 16 acres were sold to developers. The area became residential, with young families settled in the homes.

After several years of heavy rain, wastes started leaking from the buried drums. By 1958, chunks of phosphorous had migrated to the surface and were picked up by children, resulting in burns. Birth defects began to show up. Sludge and hazardous liquids surfaced in basements and swimming pools.

In the spring of 1966, two young brothers fell into a muddy ditch near their neighborhood playground. Their clothes, which were covered with oil, were disposed of, but a prevalent smell remained for three weeks. The family alerted the County Health Department of their concern. The Health Department conducted an assessment by taking samples for analysis. They reported back to the family that nothing was wrong.

Shortly after that, tests of city water showed the presence of toxic halogenated compounds. Still, no action was taken, and no warnings were issued by any level of government. In addition, air samples from basements were starting to show toxic levels well above the threshold limit values.

Figure 12.2 Love Canal in 1984. (Reprinted from U.S. EPA, "The Disposable Society," *EPA Journal*, 14/7, Washington, 1988.)

In 1978, a thorough assessment was conducted, and it was discovered that the playground was the top portion of the dumping ground. The assessment revealed 43 million pounds of approximately 80 different chemicals. Of special concern were benzene, chloroform, trichloroethylene, methylene chloride and dioxin. Based on this and other evidence, New York State health officials in 1978 advised pregnant women living in the area to move immediately. Shortly thereafter, the state bought the homes of 239 families for $10 million, forced the people to move and fenced in the area.

In 1980, President Carter ordered the evacuation of 700 families from the Love Canal area. At this point, another study had been released to the public showing evidence of chromosome damage.

Hooker Chemical Company stated many times, in its defense, that it had followed proper disposal techniques required at the time. Documentation was presented to show that their past disposal techniques, which were used more than 30 years earlier, even conformed to regulations being adopted by the Carter Administration. The company claimed that an impervious clay liner and cover were used, but that the cover may have been disturbed during grading. After much shifting of responsibility and finger-pointing, an out-of-court settlement of about $30 million was reached between 1345 residents of the Love Canal neighborhood and the parent company of Hooker Chemical, Occidental Petroleum. See Figure 12-2, which is a 1984 photograph of Love Canal [EPA 1988].

CHAPTER

THIRTEEN

Costs

Introduction

Once again, as in most other fields, the power of the dollar is felt, and stronger than ever. There are essentially unlimited costs associated with HMW management. Some are obvious, such as engineering, supplies, construction, equipment, monitoring, training, fines, remediation, permits and disposal fees. Many others are sometimes hidden, including the high cost of insurance, loss in property value, management expenses, litigation and reputation. This chapter will address some of these costs.

But first, why is it important to understand the issue of cost? It is important because the number is high, and it affects decisions that are made concerning HMW. A good manager needs to know how to get the "biggest bang for the buck" when he or she is managing HMW. It is also important for the regulator to know how much different organizations are spending on HMW management. After all, is it fair, in our competitive society, for one organization to be spending $.05 on the dollar for environmental protection while another organization of the same type is spending $.50 on the dollar? How long will the environmentally conscious company survive?

Hazardous-Material-Related Costs

Up-Front Costs for Hazardous Material Permits and Approvals

Some of the first costs that hit occur when a company orders hazardous materials. Permit fees and license requirements must be paid to a variety of agencies, depending upon the hazardous nature of the material. A permit and accompanying fees are often required by the fire department for storage and use of many hazardous materials on-site. In addition, a state license, with applicable fees, is usually necessary for radioactive materials use. If an underground storage tank is used in the operation, a permit with fees is required by either the county or the fire department. Counties also require permits and fees for use and storage of various other hazardous materials. An example of some actual underground storage tank fees include:

- Permit For Use—County; $250/year

- Storage of Hazardous Material—City/County; $ varies

- Tank Registration—State; $25 one time

- Board of Equalization—State; $200/year/tank

Construction Costs to Allow Storage and Use of Hazardous Materials

A tremendous amount of obvious and hidden costs occur once the HM is on site in terms of building considerations. Special storage conditions, at considerable cost, are usually required by the fire department for HM. Depending on the situation, these can range from a low of $200 for signs and a used flammable storage locker up to $50,000 to $100,000 for major building modifications and equipment. For example, specially constructed flammable material or toxic gas cabinets, secondary containment, and spill cleanup supplies may be required. The cost to upgrade storage tanks and modify treatment equipment can be extremely high. Construction modifications may even be needed for several rooms so that the HM can be stored and used safely.

Additional chemicals brought in may change the UFC occupancy ratings. For example, a B rating (office) may be changed to an H rating (chemical use and storage) if chemical usage or storage increases. This may mean additional engineering controls such as new ventilation systems, fire-rated rooms, fire-rated walls and extra fire sprinklers, all at extra cost.

Monitoring Costs

Monitoring costs may also be incurred with HM. Many HM have Permissible Exposure Limits (PELs) and environmental release limits. If so, the cost of monitoring these chemicals must be considered. As with building considerations, the cost range is broad. At the low end, an occasional formaldehyde test would cost $20 or $30 for materials and labor, not including the one-time cost for purchase of the sample pump. On the other end of the spectrum, an installed monitoring system for ongoing assessment of a highly toxic chemical can cost several hundred thousands of dollars. The cost of the monitoring equipment and the time spent collecting, analyzing and recording data would have to be considered. Depending on the HM and the site-specific conditions, in-place monitoring instrumentation may be the preferred option. This would result in higher capital costs but lower operating costs. A second option would be portable field-monitoring-type equipment, which would equal less capital costs but higher operating costs. The monitoring could also be done by consultants with no capital costs involved. However, this would result in higher operating costs since this type of monitoring is usually ongoing.

Cost of PPE and Training

Special protective equipment and clothing is another cost to be considered. If an HM has a PEL, it will also probably require personal protection in the form of lab coats, gloves, safety glasses, goggles, respirators, aprons and other protective measures. The cost of this equipment varies and is easy to calculate. There are hidden costs, however, which are often overlooked, such as personal training. Proper training is required for the employee in how to use the equipment, especially respirators. The trainer's time and the employees' time in class must be added into the total cost of using an HM.

Overall Management Costs

The last type of HM cost is perhaps the most difficult to quantify, that of overall management, tracking and control of HM. One reason why these costs are hard to identify is that they are often incurred by individuals who are responsible for HM and other non-HM functions. Therefore, when these people are doing their jobs, some of their actions may affect related responsibilities, such as environmental, safety, equipment protection and production. This makes it hard to separate purely HM-related time spent. Involvement with various partially related activities, such as materials management plans, methods and procedures, quality control, mass balance tracking, inventory control, business plans and many other management actions, impact or are related to the HMW arena, and all add cost.

Hazardous-Waste-Related Costs

Many of the costs discussed above for HM directly or indirectly apply to HW as well. For example, the management costs, training and personal protection costs already presented for HM would largely cover many HW issues as well. There are a few unique costs, which will be discussed below.

Up-Front Costs for Hazardous Waste Permits and Approvals

Generators must obtain EPA identification numbers and permits for disposal. Hazardous and extremely hazardous waste permits have associated fees. It also costs money in terms of employee time to prepare the permit applications and then monitor operations to make sure that permit restrictions are not violated.

Hazardous Waste Disposal Fees

Most generators have to pay fees to the state. For example, any individual that generates greater than five tons of hazardous waste/year or disposes of greater than 500 lbs. of hazardous waste/year to the land must pay fees in California. This requires registration with the State Board of Equalization.

The transporter and final disposal site charge significant fees. Table 13-1 presents some of these fees and other hazardous waste costs. Most generators have to pay the disposal, generator, environmental and miscellaneous activity fees. The facility fees are paid by the TSDF. However, the expenses are probably passed along to the generators that use the facility, at least eventually. Also, the last fee shown in Table 13-1 is only charged to the unfortunate parties involved with a Superfund cleanup. In addition, a generator may have to pay quarterly taxes, manifest fees, closure costs and tank or treatment unit certification fees.

Table 13.1 Hazardous-Waste-Related Fees (California and RCRA)

Fees	Categories	$/Ton
Disposal Fees	Non-RCRA waste (up to 5000 tons/month)	26.25
	Hazardous mining waste	
	(up to 5000 tons/month)	13.65
	Extremely hazardous waste	210.00
	Restricted hazardous waste	210.00
	RCRA waste	42.42
	Out-of-state disposal	0
	Incineration/treatment residues	5.25
	Certain surface impoundments	5.25
Generator Fees	5-25 tons/year	141.00
	25-50 tons/year	1130.00
	50-250 tons/year	2825.00
	250-500 tons/year	14,125.00
	500-1000 tons/year	28,250.00
	1000-2000 tons/year	42,375.00
	2000 or more tons/year	56,500.00
Facility Fees	Mini storage (<1000 lbs./month)	5882.00
	Mini treatment (<1000 lbs./month)	11,763.00
	Small storage (<1000 tons/month)	23,526.00
	Large storage (>1000 tons/month)	47,052.00
	Small treatment (<1000 tons/month)	47,052.00
	Large treatment (>1000 tons/month)	70,578.00
	Disposal facility	235,260.00
Environmental Fees	50-99 employees	100.00
	100-499 employees	500.00
	500 or more employees	1000.00
HSA Activity Fee	Size and complexity estimate	5884.00
(Superfund sites)	Preliminary assessments	8828.00
	Removal action oversight	17,065.00–173,005.00
	RI/FS oversight	25,305.00–235,380.00
	RAP oversight	5297.00–44,723.00
	Design oversight	8828.00–94,153.00
	Remedial action oversight	11,768.00–124,752.00
	O&M oversight	7061.00–40,016.00

Table 13.1 Hazardous-Waste-Related Fees (California and RCRA) (Continued)

Fees	Categories	$/Ton
HWCA Activity Fees (permits and variances)	Land disposal	
	Small	97,683.00
	Medium	208,311.00
Small = 0–.5 tons/ month	Large	357,778.00
	Storage/Treatment	
Medium = .5–100 tons/ month	Small	20,008.00
	Medium	36,483.00
Large = >100 tons/ month	Large	70,615.00
	TTU	
	Small	15,301.00
	Medium	35.306.00
	Large	70,615.00
	Variances	
	Small to Large	300.00–8000.00
	Permit-by-rule notice	
	Medium	1140.00
	Waste classification	
	Medium	8828.00
	Extremely hazardous waste permits	
	Medium	235.00

Source: California Environmental Protection Agency—Fees for FY 93/94.

Overall Management Costs

Hazardous waste controls, procedures, training and documentation are necessary at considerable cost. Most of these expenses are similar to the ones discussed under hazardous material tracking and control, and some may even be the same. As mentioned above, these expenses are very hard to estimate.

Fines and Penalties

Even after tracking and control, an organization may still incur costs due to regulatory noncompliance. Fines, litigation expenses and penalties, plus the costs of corrective action, have been known to bankrupt companies. Some of the possible fines and penalties are shown in Table 13-2. The largest fine shown is $250,000/day under RCRA. It is possible that a generator could be fined under several laws and have to pay to more than one agency.

Table 13.2   Examples of Penalties

Law	Examples of Penalties per Violation
Resource Conservation and Recovery Act	$250,000.00/day and/or five years imprisonment
Hazardous Waste Control Law	$50,000/day and three years imprisonment
Clean Water Act	$25,000/day and one year imprisonment
Hazardous Materials Transportation Act	$25,000/day and/or five years imprisonment
Toxic Substances Control Act	$25,000/day and/or one year imprisonment
Porter-Cologne Water Quality Control Act	$50,000/day and/or two years imprisonment
EPA-Superfund	Liability for all cleanup and damage cost
California-Superfund	$25,000/day and/or one year imprisonment
Prop. 65	$250,000/day and six years imprisonment. Has a "bounty hunter" provision that gives 25% of the fine to the informant
Community-Right-to-Know (AB2185)	$25,000/day and/or one year imprisonment

Cost of Bad Publicity

The last cost associated with hazardous waste has also caused the demise of more than one organization in the past. This is the cost of bad publicity and poor community relations. An incident, fine, perceived poor environmental record or other action (or lack of action) can hurt an organization right where it counts, in the pocketbook. Environmental problems, especially hazardous waste, have caused customers to stop or reduce buying a product or service.

Remediation Costs

Table 13-3 presents actual bottom-line remediation costs encountered at 31 completed Superfund cleanup sites. Since the study was published in October 1987, the dollars would have to be inflated to present dollars. To be on the safe side, one would probably want to add 4% to 10%/ year from 1986 (when the report was being prepared) to the present year. The costs are divided into seven different categories or types of cost. Within each category, the unit cost values have different measurement units. These unit costs vary significantly over time depending upon availability of materials and many other factors.

Table 13.3 Range of Unit Costs Associated with Remedial Technologies

Technology	Expenditures	Range of Unit Costs[a] Estimates
Surface Controls		
Surface sealing	$0.92 to $15.84/yard^2	$1.32 to $16.88/yard^2
Grading	N/A	$4000 to $16,205/acre
Drainage ditches	N/A	$1.27 to $6.04/linear foot
Revegetation	N/A	$1214 to $8000/acre
Ground Water and Leachate Controls		
Slurry wall	$0.25 to $31.96/foot^2	$4.50 to $13.88/foot^2
Grout curtain	$6.60 to $14.00/foot^2	$5.50 to $75.52/foot^2
Sheet piling	N/A	$8.02 to $17.03/foot^2
Bottom sealing by grout	N/A	$9.00 to $116.00/foot^2
Permeable treatment beds	N/A	$14.00 to $267.00/foot^2
Well point system	N/A	$803.00 to $8284.00/well
Deep well system	N/A	$4862.00 to $13,513/well
Extraction/injection well system	N/A	$37.50/vertical foot
Extraction wells/seepage basin	$31,269.00/sytem	$33,618.00 to $53,360.00/system
Subsurface drain	$24.00 to $1733.00/foot	$1.94 to $218.00/foot
Aqueous and Solids Treatment		
Activated sludge	$6.3 million/mgd	$200,000.00 to $390,000.00/mgd
Lagoons	N/A	$80,000.00 to $3.4 million/mgd
Rotating biological contractors	N/A	$0.9 million to $29.6 million/mgd
Air stripping	$0.10 to $0.40/gallon	$14,132.00 to $643,000.00/mgd
Oil/water separator	$289,200.00/system	$12,720.00/mgd
Gas Migration Control		
Pipe vents	N/A	$445.00 to $1310.00/vent

Table 13.3 Range of Unit Costs Associated with Remedial Technologies (Continued)

Technology	Expenditures	Range of Unit Costs[a] Estimates
Trench vents	N/A	$35.00 to $646.00/linear foot
Gas barriers	N/A	$0.39 to $3.00/foot^2
Carbon adsorption	$188.00/filter	$635.00/filter
Material Removal		
Excavation, transport and disposal	$4.70 to $884.00/yard^3	$379.00 to $434/yard^3
Hydraulic dredging	N/A	$1.25 to $3.54/yard^3
Mechanical dredging	N/A	$1.37 to $4.09/yard^3
Drum handling	$60.00 to $1528.00/drum	N/A
Water and Sewer Line Rehabilitation		
Sewer line replacement	N/A	$53.90 to $141.60 /linear foot
Sewer line repair/cleaning	$15.00/linear foot	$5.75 to $15.90/linear foot
Water line repair	N/A	$26.00 to $35.50/linear foot
Water main replacement	N/A	$58.50 to $119.18/linear foot
Alternative Water Supplies		
New water supply wells	N/A	$46.25/linear foot
Water distribution system	$1091.00 to $10,714.00/house	N/A

[a] Unit costs are *exclusive* of operations and maintenance costs. In some cases, no range is given since only a single data source was available.
N/A = not available
(Reprinted from EPA, *Compendium of Costs of Remedial Technologies at Hazardous Waste Sites*, EPA Report No. 600/2-87/087, Washington, 1987.)

Case Study

Research Laboratory Using Radioactive Materials

Hazardous material and waste costs associated with setting up a laboratory will be given. The costs will pertain to only one of many waste streams. The research laboratory has 70 people in it.

If radioactive materials are to be used in the laboratory, an application for a license to the state must be submitted. The application must include a radiation safety manual for the company. The manual includes protocols, emergency procedures, equipment, personnel resumes and safety committee names and functions. Much preparation is necessary prior to submitting the application. In one situation, approximately 115 hours were spent to complete all the necessary documents for the license application. The license costs $375/year. The fire department also requires $135/year for its own type of permit.

A generator permit is required in the state where disposal occurs. In this case, the State of Washington charged $175/year. Time is also spent to obtain a generator number and to prepare the permit application itself.

Any person who may possibly be exposed to radiation above the permissible limit must be monitored. In this example, 43 different individuals required monitoring at a cost of $100/month. Some radionuclides require thyroid scans and urinalysis at a cost of $50/survey. This does not include at least two hours of work missed by each individual while traveling to and from the laboratories.

Environmental monitoring is also required. This is done on a per-use and per-week basis. The approximate time expenditure is three hours of time per week. The environmental monitoring equipment costs $6000.00 and requires a calibration every six months at a cost of $150.00

Training and informal informational meetings are also required. The training ranges from $50 to $100/person not including the 80 hours of work time spent by each trainee. The informational meetings are held quarterly and result in at least two to four hours/person.

Special equipment and building modifications are required to segregate and restrict access. This requires the purchase of three new freezers and a refrigerator for the radioactive materials. New walls and ventilation result in at least $1100.

Depending upon the type of waste, the disposal costs are approximately $700/55-gallon barrel. Four barrels/quarter are generated of the most common type of waste. At least one hour/barrel was spent tracking and documenting this one type of waste.

Overall there were 293 total man hours used/year at a projected overhead cost of $40/hour. When the permit, training and disposal costs are added in, the total of $28,555/year is reached. This figure does not include many expenses, such as one-time costs for problems, fines, litigation, and so on. When one considers that this is a low-end figure and applies only to one waste stream, it becomes obvious that hazardous material and waste expenses are high [Wert 1990].

CHAPTER

FOURTEEN

Waste Minimization

Introduction

There are few topics in HMW management on which all the players agree. With the great variety in background, perspective and purpose, this is not surprising. The one and possibly only subject on which everyone, from the most hardened industrialist to the most radical environmentalist, seems to reach consensus is waste minimization. Waste minimization is the reduction, to the maximum extent possible, of hazardous waste that is generated, treated, stored or sent for disposal.

The EPA has put waste minimization as a top priority in terms of hazardous waste management. Figure 14-1 shows waste minimization as the most preferred option. Resource recovery is shown second and really is part of waste minimization, as is treatment, but to a lesser degree.

From a global perspective, waste minimization is a common theme, and in many respects the Far East and Europe lead the United States. This may in part be due to a shortage of space and resources in those areas of the world. Therefore, there has been a long-standing practical reason to minimize waste. Without question, the Far East and Europe lead the world in nonhazardous waste minimization (such as the German packaging and product take-back regulations) and in many aspects of hazardous waste minimization as well.

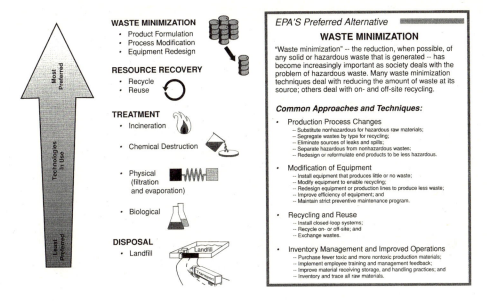

Figure 14.1 Hazardous Waste Management: What's the Best Approach? (Reprinted from EPA, 1993.)

Waste minimization is known by various names, such as waste prevention/avoidance/reduction, pollution prevention and other terms. For the purposes of this book these terms are considered similar, even though there are some differences.

Waste minimization, whether it be for industry, households or any other type of generator, is the way of the future. Land bans and increasing treatment and disposal costs make waste minimization of paramount importance. This chapter will discuss the benefits of waste minimization, priorities, key elements, ranking, options and assistance. The emphasis is on HMW; however, most of the concepts could also apply to nonhazardous material waste minimization.

Benefits of Waste Minimization

Reduced Environmental Impact

Waste minimization results in reduced environmental impact since it reduces waste in the environment. This is of great importance because the benefit is felt by almost all components of the entire ecosystem. If humans exercise waste minimization with energy and dedication, it may prevent the earth from suffocating in all the wastes that humans generate.

Improved Employee Safety

There will be fewer hazardous materials and wastes for employees to handle in an organization if waste minimization is practiced. Reduced handling equates to less exposure and improved employee safety.

Regulation Compliance

Waste minimization practices allow compliance with several federal regulations. To comply with RCRA, the generator must certify that they are implementing waste minimization. There is some waste minimization reporting associated with SARA Title III. Biennial reports require a description of the efforts undertaken during the year to reduce the volume and toxicity of waste generated. Waste minimization will help organizations achieve compliance with federal and state land disposal bans. Finally, TSDF permits require that the generator of the hazardous waste have a program in place to reduce the volume or quantity and toxicity.

Two more recent important federal bills include the Wolpe Schneider Bill, or Waste Reduction Act of 1989 (HR 145), and the Lautenberg Bill, or Pollution Prevention Act of 1990. To comply with this latter act, all facilities are required to submit, with their Form Rs, an annual toxic chemical source reduction and recycling report. The report must include: the amount of wastes entering the waste stream and percentage change from the previous year; the amount of wastes into the waste stream expected to be reported for the next two calendar years; the amount recycled, percentage change in amount recycled from the previous year and process used; source reduction practices used; ratio of the reporting year's production to the previous year's production; techniques used to identify source reduction opportunities; and the amount of toxic chemical released as a result of one-time events.

California Senate Bill 14 is resulting in dramatic changes in companies and agencies concerning waste minimization, especially source reduction. As of September 1, 1991, all generators of 12,000 kg/yr or more of hazardous waste or 12 kg/yr of extremely hazardous waste have to submit several reports. In order to prepare the reports, new organizational programs were needed in most companies. These reports include Source Reduction Evaluation Review And Plan, Source Reduction Evaluation Review And Plan Summary, Hazardous Waste Management Performance

Report, and Hazardous Waste Management Performance Report Summary. California SB 1726 in 1992 broadened the hazardous waste minimization requirements and included more small-quantity generators.

For each hazardous waste stream identified, the plan must include an estimate of the quantity generated and an evaluation of the source reduction methods available to the generator. In addition, the plan must include reasons for the reduction methods selected and a full explanation for the rejection of methods. Last, an evaluation of the effects of the chosen method and a timetable for adoption must be included.

Operating Cost Reduction

After some possible up-front expenditures, waste minimization usually results in operational cost savings. The greatest savings will be in less HMW storage, transportation and disposal costs. This is significant since disposal costs have been increasing by as much as 50%/year. There is also cost savings due to less raw materials used and less manpower needed (smaller legal, regulatory and environmental staffs, for example). Finally, there may be tax advantages and lower insurance costs.

Improved Public Relations

Improved public relations is a big benefit of waste minimization that's hard to quantify. Public relations has been known to make or break an organization, especially in the hazardous waste arena. 3M and Dow Chemical are two examples of companies that have received considerable positive recognition for their programs.

Most waste minimization efforts can lead toward Green Certifications, which are becoming popular in many parts of the world. Europe leads with programs such as the Green Point, Eco Mark, Blue Angel and White Swan. The U.S. has the Green Cross and Green Label programs. All of these programs are voluntary and oriented toward marketing, public relations and waste minimization.

Reduced Liabilities

Waste minimization can help to minimize future cleanups, court action and fines for noncompliance. As with public relations, this benefit is also hard to quantify and can be in the multimillions of dollars [Higgins

1989]. If there are fewer HMWs present, there is less chance of problems leading to fines and law suits.

Generic Waste Minimization Program Key Elements

The following list presents some of the major steps in setting up a waste minimization program. Depending on the number and complexity of waste streams, fewer or additional steps may be required. Large waste minimization projects should be carried out by a team. The team needs to be identified early, at least by step 5. If a team is utilized, it is essential to assign responsibilities. Steps 2, 4, 9, 10 and 13 are especially important and are designated by the EPA as "Effective Program Elements."

1. Set and Prioritize Objectives—Take care to clearly state a manageable number of realistic objectives. Most establish too many goals and then "shotgun" the team's energy in too many directions. This can lead to none of the objectives being accomplished.

2. Perform an Initial Waste Stream Reduction Audit—The audit should include the types, amounts and level of wastes generated and the sources of the wastes. In other words, don't set everything up from a conference room or a desk. Physically get out to where the wastes are being generated and learn about their evolution.

3. Identify and Prioritize Waste Streams to Minimize—The prioritization should consider what is feasible and what is not. In identifying the waste streams to pursue, involve all employees, not just environmental compliance staff. For example, include the workers and managers involved in production and maintenance.

4. Obtain Top Management Support—The president of the organization must support waste minimization verbally, in writing and financially. Without this support the program is doomed. It is also a good idea to get a show of support from the senior vice presidents as well.

5. Perform Periodic On-Site Assessments—By routinely going out in the field and reviewing operations, new waste streams will become obvious candidates for minimization.

6. Involve, Motivate and Educate All Employees—In addition to including as many people in the identification phase as possible, the entire organization should be involved in most other phases of waste minimization. For example, reward programs will encourage employees to think and support new ideas. Many

companies have found that dollar awards, plaques, and honorable mentions in company newsletters are all good ways to increase employee involvement [Higgins 1989].

7. Design and Evaluation of Action Plans—An action plan needs to be carefully prepared. This may even include some design. A technical and economic evaluation of the action plans would then occur and result in the selection of a manageable number of realistic action plans to be implemented.

8. Testing of Selected Action Plans—A test run of the action plan(s) is advisable. It's easier to get the bugs out on a small scale.

9. Obtain Funding—Many good projects have come to a screeching halt because of a lack of funds. The funds should be firmly committed before much energy is expended.

10. Revise Accounting Methods and Distribute—Certain waste minimization practices will require changes in the organization's accounting system. If this is identified as being needed, it should be started early, since most accounting systems take considerable time to change. Accounting systems must be in place to allow for the different ways in which the hazardous materials and wastes will be handled and tracked from a financial standpoint.

11. Revise Operational Procedures and Distribute—As with accounting, sometimes methods and procedures will need to be written or rewritten to include the new waste minimization components. They must be written clearly and in terms that all affected individuals can understand. Many refer to these instructions as "best management procedures or practices."

12. Implement the Waste Minimization Actions—This is what it's all been building up to and what all the planning is about. Now it's time to physically start the new system or process.

13. Evaluation Program and Follow-Up—It is important to ensure that the waste minimization techniques are being utilized since old habits are sometimes hard to break. Follow-up visits and memos will help ensure this.

In July 1988, the EPA published its Waste Minimization Opportunity Assessment Manual. In this document, the EPA basically identified four steps: 1. Planning and organization (see steps 1–4 above); 2. Assessment phase (see step 5); 3. Feasibility analysis phase (see steps 7 and 8); 4. Implementation (see step 12).

Waste Stream Ranking Considerations

Some operations can involve hundreds of hazardous waste streams. It may be hard to decide which one(s) to concentrate on first. The following considerations may help in this determination and are arranged roughly in order of importance, depending on your perspective: environmental and employee safety impact, feasibility of implementation, regulatory compliance (land bans), effectiveness of the waste minimization action, disposal costs, production and quality risk, monthly amount produced, operating and capital costs, and investment potential [Higgins 1989]. Volume of waste generated and exposure time are indirectly covered in the above categories.

The environment and employee safety impact component mentioned above can itself be subdivided into several categories. One important component is a toxicity rating, which commonly uses LD50. Closely related is the waste's carcinogenicity, mutagenicity or teratogenicity. The waste's flash point and reactivity are also important considerations.

Waste Minimization Options and Priorities

The waste minimization options could in general be applied at the manufacturer, distributor, storeroom, user and contractor's facilities. A few of the options may, however, be a little more practical at one phase of the HMW flow than another. Most of these options could also be applied to industry, households and the other types of generators. The following options are presented in order of priority and grouped into either hazardous material minimization or into hazardous waste reduction categories.

Hazardous Material Minimization (Source Reduction)

The heart of HM minimization is source reduction. If possible, the HM should not be allowed to enter the operation in the first place; otherwise, minimize the amount needed. HM minimization can be achieved through the following overlapping options.

Substitution   If at all possible, replace products or raw materials that contain HM for those that don't. If that doesn't work, replace with ones that contain less HM in terms of volume or toxicity. Substitute processes and equipment that use HM for those that do not [Higgins 1989].

New process or equipment substitution is the most desirable form of waste minimization. You simply design a new system, process or piece of equipment that does not contain or use hazardous materials. This is usually harder than it sounds, but definitely achievable.

Existing process or equipment substitution is the next best thing to do. Some of the HM may already be in the system. By making substitutions, changes and upgrades you can possibly phase out the HM or at least minimize it. There is usually some resistance from operational personnel when this approach is taken since they are already dependent on the HM. Sensitivity and planning are therefore necessary.

Figure 14-2 is an example of the process taken by the author to introduce a lesser toxic industrial adhesive. The first half of the figure illustrates the steps used to select the hazardous material to minimize. In this case, it turned out to be an industrial adhesive. The second half of the figure illustrates the screening process used to select the best substitute for the presently used industrial adhesive.

Inventory control   By buying only what is needed at the time and rotating, the inventory of HM can be kept at a minimum. Also, if the number of different brand names can be reduced, it will help keep the inventory at a lower level.

Purification of raw materials   Sometimes, if one or more of the raw materials can be purified before being fed into the process, it will generate less waste. The manufacturer should be encouraged to clean up its product to higher quality standards. The purchaser may still need to purify the raw material even further. In addition to reducing hazardous waste, this may even improve the quality of your own product or service.

Development of new operating procedures   It is possible to write procedures that will help minimize HMW. For example, an operating practice to screen MSDSs for all new products purchased or about to be purchased would probably help minimize the amount of HMW the organization must contend with. These procedures are also known as management practices, standard operating procedures and other terms.

INVENTORY WASTE STREAMS

EVALUATE AND PRIORITIZE

SELECT TARGET WASTE TO ELIMINATE

SCREEN POSSIBLE SUBSTITUTES (COMPUTER AIDED)

QUALITY AND COST SCREENING OF SUBSTITUTES

ENVIRONMENTAL AND SAFETY SCREENING OF SUBSTITUTES

INTRODUCE SUBSTITUTES

FOLLOW-UP

Figure 14.2 Waste Minimization Process Example.

Production scheduling Tightening up the production schedule, so that there are fewer start-ups and shutdowns will minimize waste. This is possible because more waste is usually generated when the machines (and operators) are cold. Alternating operating schedule and rearranging work shifts might minimize cumulative product waste, emissions and risks.

Better housekeeping Housekeeping was also discussed in Chapter 9. Keeping areas clean, in order and with a minimum of waste present is an excellent way to facilitate all of the other waste minimization options. Labeling and segregating HMW are also good housekeeping practices.

Hazardous material reuse, exchange or sale Some HM can be reused within the same organization, even without treatment. For example, paint thinner can be reused many times, merely by allowing a few hours of settling time. Also, many high-grade solvents used in an ultraclean manner may be reused in areas with less stringent quality standards.

If some effort is expended, it may also be possible to find someone outside of the organization who may want to have the HW. Waste exchanges help facilitate this by listing unneeded HMW from various organizations. Waste exchanges were developed in Europe and are starting to become established in the United States. If the HMW is exchanged, donated or sold, it might not even be considered a waste since it has a functional use in its present form. It is recommended that when this occurs, a bill of sale is prepared, even if it is for $1. When something is sold, considerable responsibility (but not all) passes to the new owner.

Improved efficiency If the entire flow of HM is made as efficient as possible, HM can be minimized. Efficient manufacturing, distribution, storage and use will aid in the effort. Manufacturing and materials handling are two areas where efficiency improvements will have a big impact.

Better equipment maintenance and monitoring Maintenance and monitoring are closely related to improved efficiency. If equipment utilizes or produces HMW, it should be monitored and maintained. This will improve the quality and minimize the amount needed. Improved maintenance will help in reduction of spills, leaks and vaporization.

Improved mass balance tracking and product conservation Also related to the above is mass balance tracking. Complicated processes may have interrelated hazardous material streams. It is essential to accurately account for all of the material at all times. Among other things, this will help conserve the raw materials.

Altering the process By adjusting the process it is possible to eliminate or reduce hazardous material end products and byproducts. This is closely related to the section already presented, "Development of new operating procedures" on page 223. For example, using physical or mechanical cleaning (such as hydroblasting, rodding, brushing, wiping and scraping) in conjunction with solvents will reduce HW. These are modifications that result in less chemical use or safer chemicals used.

Hazardous Waste Reduction

If it is not possible to further minimize or reduce the hazardous materials at the source, HW reduction should be implemented. At this point, we are concentrating on reducing the total volume or toxicity of waste that has already been generated.

Recycling, reclaiming and reuse Certain processes allow for reuse of part of the waste stream after some treatment. For example, organic solvents, antifreeze, silver, sulfuric acid and oil are commonly recycled. If possible, on-site recycling is preferred over off-site.

Assuming you can't minimize the HM at the source, reuse and recycling is the next best option. This may require permits and still generate some waste, but at least part of the waste stream will be reutilized, thus minimizing the amount of new HM needed and the amount requiring disposal.

Recycling is one of the easiest options to recognize. When most individuals refer to waste minimization, they are normally using recycling examples. Waste oil, antifreeze and batteries are recycled by many people. Most industries are now recycling solvents. Specific industries are even recycling unique waste materials, such as waste sulfuric acid.

Most recycling processes have some economic limitations on the type and quantities that can be recycled. The market for some recycled materials is acceptable, while other markets are flooded. This fact determines how much energy and dollars can be spent to recycle different wastes. The cost of straight disposal is also a consideration.

Recycling is done either on-site or off-site. If the volumes allow it to be done economically on-site, transportation costs can be avoided. Sometimes a TSDF permit is required whenever any new hazardous waste is created from the recycling process. In some states there are some exceptions, variances and alternate permit options available that avoid the requirement for a full-blown TSDF permit. Also, if the hazardous material is pulled out of the process and recycled before it becomes a waste, and the recycle process does not generate a hazardous waste, a TSDF permit may not be required. In this situation, it would be considered a closed loop.

Recycling in many cases basically involves the removal of impurities and the addition of additives to functionally restore the material. The removal step involves processes that separate dirt, metals, bacteria and other contaminants from the product. Examples of removal processes include settling, filtration, straining, cyclone separation, centrifugation, magnetic separation, pasteurization/distillation and ultrafiltration. Once the contaminant is removed, the material is used as is or functionally restored by adding conditioners, emulsifiers, concentrates, surfactants, bactericides, antioxidants and other chemicals.

Many factors must be considered when deciding whether to recycle and, if so, with what system. Probably at the top of the list are the economic factors such as present disposal cost, and type and quantity of waste.

Equipment effectiveness, operational simplicity, maintenance, floor space requirements and, of course, degree of environmental impact reduction are also important [Higgins 1989].

Improved waste tracking As with HM tracking, it is essential to have an accurate mass balance for HW. The manifest system is an integral part. However, prior to manifest preparation, the generator can help minimize waste by better waste tracking. This would involve knowing where all effluents, emissions, bulk and containerized waste are sent.

Waste as a fuel or construction material In certain cases and with agency approval, some HW can be used as a fuel (such as in cogeneration for energy recovery), as a construction material additive (such as in concrete) and for other uses. Some states consider the burning of used oil as recycling. In terms of construction additives, 40 CFR 248 requires that federal agencies procure building insulation materials that contain recovered materials, and 40 CFR 249 requires that concrete products contain a minimum amount of fly ash. Care must be taken, however, to ensure that the waste is adequately "fixed" and will not leach toxics into the environment if it is used as a construction material. If it is used for energy recovery, it is especially important to make sure the contaminants are completely destroyed during the process.

Treatment Treatment will minimize the volume or toxicity of the waste requiring disposal. Unfortunately, treatment often requires numerous permits, which is usually expensive. Treatment should only be considered if the other waste minimization options are completely unacceptable.

Waste Minimization Assistance

The EPA and many state agencies are offering technical assistance in waste minimization. For example, information sources, grants and loans are available for waste reduction. In addition, the California Waste Exchange promotes waste reduction and exchange.

Case Study

7-Up

Save some 7-Up for your planaria. Planaria are normally preserved with a mercury-based preservative prior to microscopic study to prevent them from curling up. 7-Up works as well in keeping the worms from curling up. This waste minimization action saved money and reduced hazard-ous waste. Who knows, planaria may be used in the next 7-Up ad campaign [Wedin 1992].

FIFTEEN

General Management Systems for Tracking and Control

Introduction

It is essential that the flow of HMW be well understood and organized. This helps to minimize impacts and accidents. Overkill is recommended until HMW are well understood and errors essentially non-existent.

Several project management tools are available and can help in the control process. The Project Evaluation and Review Technique (PERT) process commonly used for major construction projects is sometimes useful for tracking the progress of interrelated HMW tasks. A critical path emerges from the PERT, which highlights essential tasks and the time element. There are various PERT systems on the market, such as STORM (Doubleday Publications), Timeline, Premavera and Harvard Project Manager (Harvard Software).

Integration of the many programs into a meaningful system, which can be tracked and controlled, is a challenging and important goal. An overview of such an attempt is shown in Figure 15-1. This figure mentions many items presented in earlier chapters, especially Chapter 2. The whole purpose of the matrix is to allow programs prepared for one regulation to be utilized to comply with other regulations. The matrix allows common elements to become obvious.

BUILDING BLOCKS OF GOVERNMENTAL COMPLIANCE PROGRAM AND BUDGET PLAN / LEGAL ENVIRONMENTAL COMPLIANCE REQUIREMENTS	BUILDING BLOCKS OF AN ENVIRONMENTAL COMPLIANCE PROGRAM										
	HAZARDOUS MATERIALS MANAGEMENT DATA				FACILITY AND PROCESS PERMIT REQUIREMENTS			WASTE HANDLING		TRAINING	
(FEDERAL/STATE/LOCAL)	HazMat Inventory	MSDSs	Emergency Plans	Waste Manifests	Facility Modifications	Process Modifications	Monitoring Requirements	Onsite T/TSDF	Offsite T/TSDF	Program Development	Program Administration
PART I HAZMAT USE REQUIREMENTS											
• INVENTORY/DISCLOSURE – SARA Title III (EPCRA); OSHA HazCom (OSHA; HSITA, Occupational Carcinogens Act); Business Plan Act (AB 2185, AB 3777) • EMERGENCY RESPONSE – Spill Notification (CERCLA, CWA, CAA, et al.; HSAA, Business Plan Act, Porter-Cologne Act, et al.) – Spill Containment and Control (CERCLA, RCRA, et al.; HSAA, HWCA, et al.) – Spill Cleanup (RCRA, CERCLA; HSAA) • WORKER SAFETY – OSHA; HSITA – Injury Prevention Program (SB 198) – UBC/UFC • POLLUTION PREVENTION – Product Substitution (Prop 65; SB 14) – Waste Minimization (SB 14) – Inventory Management – Reuse/Recycling											
PART II HAZMAT DISPOSAL REQUIREMENTS											
• AIR EMISSIONS – Criteria (CAA; Ca. CAA) – Toxics (AB 2588) – Vehicle Management (CAA; Ca. CAA) • WASTEWATER – Discharge to POTW (CWA; Porter-Cologne Act) – Discharge to Surface Waters (CWA; Porter-Cologne Act) – Stormwater Discharge (CWA) • HAZ WASTE – Onsite Handling (RCRA, HSWA, OSHA; HWCL) – Onsite TSDF (RCRA, HSWA, SARA, OSHA; HWCL) – Offsite Transportation (HazMat Transportation Act; HWCL, Ca. Vehicle Code C6) – Offsite TSDF Use (RCRA, HSWA; HWCL) • SOLID WASTE (RCRA, HSWA, Ca. HWCL) • RADIOACTIVE WASTE (Ca. Radioactive Waste Control Act) • MEDICAL WASTE (AB 1641, AB 109)											
PART III REGULATED BUILDING COMPONENTS											
• ASBESTOS (OSHA, CAA, UBC; Prop 65) • PCBs (TSCA; HWCL) • USTs (CERCLA, CWA; Porter-Cologne Act, AB 3560)											

Figure 15.1

Manifest Tracking

In addition to the paper flow required by the agencies, other manifest double checks are suggested. For example, a generator can ask a transporter to supply a routine list of manifest shipments. The list should be double-checked against the primary manifest file kept by the generator. If variances are noted, it is possible that the organization's field forces are not sending in all copies to the primary file.

HMW Expense Tracking

Agencies, investors, officers and others are starting to ask for HMW expenditure information. "How much has the organization spent to comply with the X regulation?" Usually, it is very hard to answer this type of question unless a cost tracking system has been established. This will also help during budgeting periods and will show the magnitude of HMW in terms most people can understand, dollars. Cost tracking will also help pinpoint potential problem areas in the organization, that is, excessively low or high costs.

Solid control over costs already incurred and reasonable projections of future costs are also a requirement for compliance with the generally accepted accounting principles and financial reporting requirements of the Securities and Exchange Commission. This is especially true if the costs were or will be large enough to be of material impact to the company's financial strength or if they were incurred as a result of an environmental agency's enforcement action [Meehan 1993].

The number of codes set up to track costs should be kept at a minimum; otherwise, resistance may be encountered from both the accounting department and the individuals spending the money. The following are possible HMW cost categories:

1. Treatment and disposal

2. Permits, insurance and fees

3. Consultants

4. Fines and penalties

5. Depreciation

6. Cleanup/remediation costs

7. Court settlements and litigation expenses

8. Equipment

9. Internal labor

10. Materials, fuel, electricity and supplies.

Standard Operating Procedures

An essential component of hazardous material and waste management is the existence of standard operating procedures (SOPs) which specify how to achieve regulatory compliance. The SOPs should summarize the important regulations, in layperson terms, as they apply to the specific organization or industry. For example, there might be an SOP for storage of hazardous waste and a separate one for labeling, and so on. One high-tech electronics company has approximately 50 different SOPs, which are continually revised to adjust for new regulations. The SOPs must be user friendly and provide only information that is necessary. The SOPs, if they are written at a central or corporate level, should also allow more stringent local regulations to take precedence.

Procurement Control Systems

Because of the highly technical nature of HMW products and services, vendors may need to spend considerable time and money preparing good proposals or quotes. Due to this fact, vendors who are not awarded the business may be bitter in some cases and occasionally question the procurement procedures used. Procurement-related law suits are also not unheard of in the highly litigious U.S. society. The author was in fact named as a defendant in a multimillion dollar procurement-related law suit, which was ruled in the author's favor. The procurement controls presented in this section were practiced in real life and were the reason for the successful outcome of the law suit.

In most organizations, the environmental or health professional is assisted with HMW procurement-related activities by several other departments. For example, the procurement or purchasing department in most organizations plays an important part in the HMW game. Its involvement will help ensure the use of qualified vendors or quality products. It will also help minimize the chance of a law suit. HMW procurement activities are generally grouped into the five phases discussed below. This could apply to the purchase of HMW services or products containing HM.

Vendor Qualification

By sending out a Request for Information (RFI), visiting the vendors' facilities and reviewing their technical and business practices and procedures, a sense of their degree of qualification or ability to do a good job can be made. As many as 50 vendors may be screened, with only five to ten possessing the technical and business qualifications needed. An approved vendor list is usually prepared during this phase. The screening criteria used to arrive at the approved list should be kept in the file. Be sure you would want to do business with everyone on the qualified list, because it is very difficult to disqualify a low quoter.

Request for Quote (RFQ)/Request for Proposal (RFP)

The second major step in the procurement process is to collect the specifications needed for a particular job or product and package them into an RFQ or RFP. A suggested or draft agreement and quote procedures are included in the package, in addition to the specifications.

If the service or product is clearly understood, an RFQ can be requested, and primarily only a quote comes back from the vendor. If there is a lack of clarity or understanding, an RFP should be sent, and then the vendor sends a quote and information about how they would tackle the problem. During this phase the vendors are usually allowed a walk-through or a chance to see the situation and ask questions. All qualified vendors should be invited to the same walk-through and have an identical opportunity to ask questions and hear all answers. They must also be given the same amount of time, down to the minute, to submit a bid, quote and/or proposal.

Especially for hazardous waste services, it is usually a good idea for the RFP to contain a disclaimer of liability. The specific language should be worked out with the company attorney, but generally a disclaimer for any failure to handle the RFP in a fair manner is included. It should disclaim any liability for costs or losses incurred by the quoter. Last, it should allow the purchasing company to award the contract to anyone or no one as it may choose at its sole discretion [Meehan 1993].

Selection and Notification

There needs to be clear, documented evaluation and award criteria. For example, price, quality and demonstration of understanding should be considered in the selection process. If products are being purchased, the presence of hazardous materials should be weighted negatively. Once the selection has been made, all participating vendors should be notified. Confidentiality should also be maintained with regard to the vendor quotes and proposals.

Agreements

Negotiation of the final and comprehensive agreement is the next major phase. The agreement is especially important for hazardous services or products, and liability/indemnification issues should be spelled out in detail in the agreement.

Three types of agreements are used in terms of hazardous services or products. Confidentiality agreements may need to be signed early in the process. The one-time job agreement is usually the second type of agreement to be signed. A master agreement is the third type, and allows numerous jobs or product orders to be covered under one agreement. Both the purchasing/procurement and legal departments would be involved in these activities.

Ongoing Vendor Interface

A good long-term working relationship with several vendors is recommended. If numerous jobs are planned, work orders or purchase orders should be implemented in accordance with a master agreement. The qualified vendors under contract should be listed as such in an ongoing communication medium, whether it is a hazardous service or a product containing a hazardous material.

Chemical Database/Material Safety Data Sheet (MSDS) Control Systems

Chemical information, such as from a MSDS, can be input into computer systems. This allows anyone in the field to access important chemical information. In addition, various lists of regulated materials (such as SARA III) can be checked against the chemical file and a match list generated.

The accuracy of an MSDS should be continually questioned, tracked and controlled. It is not uncommon to find an MSDS sheet, especially from certain foreign manufacturers, with some of the chemicals not identified. Many times, when wastes are tested and profiled, a chemical will show up that is not supposed to be present. If the process didn't contribute the chemical in some way, there is a strong possibility that the chemical was in one of the raw materials, whether the MSDS showed it or not. A disclosure agreement may be needed with the chemical manufacturer in order to obtain information about some trade-secret chemicals.

Risk Management Controls and Considerations

In some organizations there is a separate risk management department. In others, it is a management control function that is performed by environmental, safety and legal professionals involved with HMW. A good portion of the topics presented in this book would be considered in the risk management decision.

Risk management involves making a decision based on the information and then taking some action, if needed, before a crisis occurs. It is managing not reacting. Risk management uses the following types of information to make an informed decision: risk assessment findings, control options that reduce risks to an acceptable level, politics and public perception, liability assessment, damage to company reputation, loss of time, costs of cleanup and insurance availability.

The available information is in. Now it's time to make a decision. It may be a decision to set new regulatory limits and standards. Possibly, it will be implementation of a control option to reduce or eliminate a chemical. In some cases, the correct decision is for no action.

When a decision is made it is also important to put the risk in question in proper perspective. For example, the lifetime risk that someone will die by smoking is one person for every 12 people, by motor vehicle accident is one person for every 65 people. For home accidents it is 1/130. Deaths from most chemical exposures range from 1/100,000 up to 1/1 million.

Case Study

Yucca Mountain, Nevada

This case study involves tracking and managing hazardous wastes for the next 10,000 years. Yucca Mountain, Nevada, is a proposed site for high-level nuclear waste, which has a half-life greater than 10,000 years. $6 billion is being spent to assess this site, including how to maintain records and markers that can be read for the next 10,000 years to prevent disturbance of the area. Will mankind be around then, and what language will people read? [*Nuclear Garbage* 1992]

C H A P T E R

SIXTEEN

The Human Element

Introduction

With the technical aspects of HMW covered, we now turn our attention to a topic that is just as important, but ignored in many references: the human and organizational factors. The topics in this chapter, more than any other, must be lived and experienced to be appreciated.

Many individuals prepare themselves to the highest degree technically for HMW careers, but have significant problems in real life because they have not considered or have underestimated the human and organizational elements. HMW is an emotional issue for a large percent of the population. Many people will have strong opinions and fears which must be considered. This discussion will start with the individual and expand to the internal organization and still further to the external population.

Most of the information in this chapter is based upon the author's observations and opinions. There are very few references that specifically deal with the human and organizational issues confronting the HMW professional.

The Individual

Ongoing Education and Training

Continued learning is an individual issue of importance, especially in this field. Since regulations and technology change so quickly most individuals need to continue learning and must do so to stay on top of the HMW field. Seminars, conventions, in-house training, short courses, hazardous certification programs and formal education help to keep the individual up to date. Large-quantity generators are required to have an established training program; however, small-quantity generators are not.

In-house employee training can take many forms, including on-the-job, videos and classes. One company utilizes a combination of two-day, half-day or one-hour HMW hazardous training, which includes four videos. The degree of HMW involvement dictates the type and extent of training. For example, all employees should have at least one hour of training. Employees who may encounter HMW should have from one-half day to two days. The environmental/safety staff should have considerable training. If the facility uses its own fire brigade, it should have significant HMW training incorporated into ER training. All employees who handle or deal with HMW must have 24 to 48 hours of HAZWOPER training.

Some specific examples of training needed for people dealing with hazardous material and waste includes asbestos, bloodborne pathogens, chemical handling, compressed gas safety, CPR/first aide, liability training, hazard communication, hazardous waste management, emergency response, hazardous waste treatment, spill cleanup, manager/supervisor environmental/health/safety training, respiratory protection and others. The majority of these classes are required by OSHA and EPA. All employees would not need to attend all of these classes, only those appropriate for their jobs. These classes provide the specific training needed so that employees handle issues quickly, safely and properly. The classes are also valuable in increasing the general awareness of all employees, which is essential for proper hazardous material and waste management.

Turmoil in the Environmental/Hazardous Field

The individual who selects a career in the HMW field can expect a great amount of change, uncertainty and turmoil. This is good for some people, especially those who thrive on change. It is not so good for those individuals who need stability and routine in their lives. Table 16-1 lists some of the causes of turmoil and possible effects on a HMW career.

It may be hard for a person to reduce the root causes of the turmoil, but it at least helps to understand that there is considerable change in this field, and the causes are very nebulous. For example, there is very little one person can do about the economy, political climate and young age of the field. A politician or organization may have a better chance to initiate positive change to reduce the turmoil in the field.

Table 16.1   Turmoil in the Environmental/Hazardous Field

Causes

- *Changes in economy*
- *Politicians*
- *Changes in attitudes (core environmental issues)*
- *Young age of the field*
- *Lack of coordination*
- *Emotional subject*
- *High liability*
- *Great amount of uncertainty*

Effects on Career

- *Interesting and stimulating (always something new to learn)*
- *New opportunities*
- *A career in flux*
- *Great turnover and movement*
- *Wide range in salaries and benefits*
- *Wide range in responsibilities*
- *Wide range in respect*
- *Career confusion*
- *Feast or Famine*
- *Conflicts between security/lower salaries and high risk/higher salaries*

In terms of effects on the HMW career, the last bullet on Table 16-1 needs additional explanation. As long as an individual is happy with a lower salary, one can usually have more security. In this field, an individual generally can't have both security and a high salary. This is the conflict that many feel. At some points in an individual's life, the high salary is important (when the big expenses are rolling in). At other times, security is more important.

Certifications/Registrations

Unfortunately, "paper" is very important to many individuals and organizations. It can be overstated, but can also serve to keep unqualified individuals out of a profession where they could do real damage. Academic degrees, while not specifically in this category, are also important, if not more so. Some of the common certifications/registrations include Professional Engineer, Certified Industrial Hygienist, Certified Hazardous Materials Manager, Registered Environmental Assessor, Certified Waste Water Treatment Operator and Registered Environmental Professional.

The Organization

The organization in this discussion could apply to any group of people involved with HMW, including an industry, home, agency, consulting firm, environmental group, university, media or legislative organization.

Structure

The mechanistic or hierarchial organizational structure in older industries is not as well adapted to handling HMW problems as are the newer, organic forms of organization. In the organic structure, less attention is paid to chain of command and more toward functions that fit the need of the organization, the client and the individual. This seems to aid quicker resolution of HMW issues.

Communication

Communication of HMW issues is one of the most sensitive and difficult functions for the HMW professional. It is also badly handled by many technical experts. The right people must be informed at the right time with the right amount of detail. The right amount of their feedback must be incorporated. It is extremely hard to get all of these "rights," in the proper amounts, when many different types of individuals are involved, each with a unique set of perspectives [Friedman 1988].

Emotions are more commonly encountered in this field than in almost any other. The professional should be ready for fear and overreaction. Therefore, communication should be done carefully and with sensitivity. Communication should also be done routinely and prior to action. Consider that the audience may have varying levels of understanding and certain prejudices with regard to particular practices or methods of handling HMW.

It is easy to have too much or too little communication. For example, before California Prop. 65 there was too little communication about toxic exposure. Following passage of Prop. 65, there has been so much communication by signs and newspaper articles that the public is starting to ignore warnings, even the important ones.

The preparation of a communication plan will help. This allows the strategy and steps to be well defined before the action occurs. The plan should consider the several types of audiences: the organization, community, media, agencies, customers, employees, competitors, environmental groups and many others.

Budgets

"There is not enough money in the budget to do your HMW project. This expense was not identified last year, so there is no money." These are common statements that most HMW professionals hear over and over. For organizations to survive in the competitive environment they must be cost-conscious and minimize expenses wherever possible. This applies to all departments in an organization, including the environmental department.

The HMW professional must make it clear which expenses are absolutely essential and which are not. There are many that cannot be avoided. There should be a nondiscretionary source of funds and a discretionary source.

It will also help if there is at least an attempt to predict HMW expenditures for the next five years. This may be hard to do and may have a low degree of accuracy, but it does alert budget people and top management to possibilities in the future. They are then not quite as surprised when actual expenses hit.

Waste minimization is one way to help reduce or control the HMW budget. By minimizing or avoiding hazardous materials in the first place, the cost of disposal after the material is used will be reduced. Other waste minimization approaches also minimize budget problems such as use of less hazardous materials and recycling.

Politics

One should be aware, if they are not already, that politics exist in this field as well as most other fields. The politics may be internal (organizational) or external (in the laws and regulations). Internally, some individuals may use hazard-related excuses to build staffs and buy equipment. This may lead to duplication of effort, turf wars and other wastes of time and money. Politics have also been known to exert artificial pressures on HMW professionals.

To be able to plan for the future, HMW professionals must be aware of the fact that politics are involved in the creation and implementation of new hazardous material and waste legislation to a greater degree than in many other subject areas. Many politicians have a sincere interest in the environment and will sponsor legislation for that reason alone. Some other politicians, however, will jump on the environmental bandwagon to obtain votes. Still other politicians may try to tie up legislation if it affects a powerful lobbying organization. Unfortunately, there are considerable dollars and careers dependent on the HMW political climate, which changes continually.

Hazardous waste management and environmental control in general are commonly regarded by many in an organization as an unnecessary burden and expense. Most do not understand the regulatory requirements. Others perceive HMW management as a necessary evil. Both of these negative perceptions have to be changed if the HMW professional is to be effective in the organization. One way to do this is via upper management support and communication programs.

Centralization

Some organizations promote centralization, while others encourage decentralization. Certain functions are better done at a central or staff location, while others are more appropriate in the field. In the author's opinion, the following hazardous functions should be centralized (in decreasing order of importance): design of HMW compliance policies, qualification of HW contractors, HW contractor agreements, budgets and budget control for HMW, and environmental compliance audits. If these five activities are done by a centralized staff, there will be less duplication of effort, cost savings, better compliance and less confusion. The actual implementation of HMW policies and day-to-day regulatory compliance should be done in the field.

Staffing

Adding new HMW professionals to an organization is an important aspect of HMW management. The individual should have a degree or considerable working experience in the field. Certifications and registrations are also good. The level of the HMW position should dictate the amount of experience and number of credentials.

Of equal importance are the individual's human relations skills and enthusiasm. If these are good, they can even offset degree and experience weaknesses. The opposite is not true. A highly degreed and experienced person with either low enthusiasm or poor human relations skills is unacceptable.

Organizational (Internal) Stakeholders

There are many individuals in an organization who play a part in HMW management. Some of their parts are merely FYI, and others are more involved; however, they all have a "stake" in the game or in the decision-making process. If one or more are left out, resistance may develop, and the hazard issue may be resolved poorly or not at all. This is a very real problem in larger organizations and one that can result in incredible waste of time and money.

In a typical organization, there are numerous departments or types of individuals commonly involved with HMW that should be folded into the process. Procurement is the first. This department establishes contracts with HMW vendors who are qualified and have been added to a qualified vendor's list. The legal department is there to advise and resolve contract problems and provide regulatory and litigation support. The environmental/safety staff is on hand to review regulations, manage HMW vendors, perform audits and set policy and guidelines. The manufacturing or client service department is usually the group with the environmental or safety problems. The facility department usually gets involved in making engineering controls to address HMW issues. The human resources department may help if administrative controls are needed to reduce impacts. Finally, the engineering department may offer technical data and advice.

For almost any issue, you will find people who are not really directly involved, but are more than willing to express their support or resistance. The resistors can kill a project and should be dealt with carefully. These people should be considered restraining forces and must be turned around if the issue is to be resolved. There will also be supporters, many of whom will not express their support unless asked. All of these people should be included in the communication program.

It is very important to consider all of the stakeholders and their differing perspectives. When this is done, a strategy can be developed to utilize their energies in a positive way and resolve the HMW issue or project. Failure to do this can kill a project or prevent resolution of even the simplest issue.

Getting the Work Done

There are several other considerations, in addition to the stakeholders just discussed, that will aid or deter issue resolution. If these are kept in mind, the work will get done more efficiently.

Project champion If one person is clearly designated as champion, project manager or coordinator for an issue and given some responsibility, the issue stands a better chance of quick resolution; there needs to be a driver. In real life, this person ends up doing the majority of the work on the HMW project.

Committees Depending on the size and complexity of an issue, a committee may have to be formed. The internal stakeholders would be good candidates for the committee. Care must be exercised so that the committee is a positive force. In far too many cases, committees only add confusion and delay. Some ways to keep the committee on track are to minimize its size and have distinct member responsibilities and action milestones. It is also important to get members from all the applicable disciplines.

Individual contributors Committees are good for setting general direction; however, most of the progress in resolving the HMW issue is made through hard work by individual contributors. Usually, this progress is not made in the committee setting. By appreciating these facts, the project manager can keep committee meeting time and individual work time in the proper proportions. By providing the right stimulation, direction and support, the number of individual contributors can be maximized.

Other resources Many resources exist inside and outside of the organization. It is surprising how many exist that the team forgets about or is unaware of until it is too late. The trick is to find them and then to turn as many as possible into individual contributors. Communication of the issue to as many people as possible will increase your number of resources.

The paper game   Paper-shuffling can get out of hand and slow down progress. In essence, many people in organizations end up doing endless editorial review of other editors' work, resulting in marginal value. Meeting minutes and progress memos, for example, may be rewritten so many times that actual issue resolution work slows down. A definite attempt to minimize needless paper work needs to be made. Technical reviewers should not make writing style changes.

The organization immune response   Inertia exists in both the living and nonliving world. The organization is no exception. Many people resist change and action, whether it is good or bad, and basically feel that if it is different, it is to be mistrusted. The project champion must anticipate this reaction and prepare a strategy to counter it before it happens. Change-management teachings present ways to offset this resistance, such as slow introduction and familiar packaging of different ideas.

External Population

Humans and organizations should remember that they are not closed systems. Activities in the outside world will have an impact on them, as they will on the outside world. Because of this, communication in both directions must remain open. If this is done, the human element and organization will be able to adjust and control their activities in a responsive way.

Sometimes it is hard to predict the reaction or impact on the outside world of an organization's proposed action. There are many stakeholders out there, each with their own sensitivities and backgrounds. If a little sampling or probing is done before implementing major actions, it may be possible to get a sense of the probable response.

The needs, backgrounds and perspectives of the different external groups should be considered, just as they were for internal groups. Understanding these issues will help the HMW manager predict and efficiently interface with each group.

Plants and Animals of the Environment

Far too often, our silent partners in the environment are not adequately considered until too late. Concern over human impact and reaction to a change is usually so great that the other living organisms become an afterthought in many cases. Even if an EIR or EIS is not required and no environmentalist groups are stressing the plants and animals, an organization should still try very hard to continually assess the impacts of their actions on the entire ecosystem.

Unfortunately, plants, animals and humans have suffered because of a lack of hazardous material and waste management. Thailand, for example, is overwhelmed by certain hazardous waste, such as sewage. This is seriously affecting plants, animals and humans. If you take a train ride through many countries in Europe, it is not uncommon to see stained soils and stunted vegetation in many industrial areas. The U.S. still has some areas where you can see oily sheen on many water bodies, for example. The answer to all this, of course, is hazardous material and waste management.

Media

Since HMW is an emotional issue, it is a favorite topic for the newspapers, TV and radio. Unfortunately, the issues are sometimes exaggerated or misquoted. By developing media contacts during non-crisis situations and maintaining them, the possibility of sensational journalism will be kept partially under control. The media needs the name of a knowledgeable individual in an organization who they can contact without being transferred over and over. This first point of contact should be made known to the entire organization before an accident occurs, and the contact's number should be posted or otherwise readily available.

The contact person named should be thoroughly coached by an individual knowledgeable in media relations. For example, the contact needs to know that if a reporter asks about a HMW problem, he or she should answer the question in a true, brief and positive way. It is best to never guess at an answer; rather, say you do not know. If at all possible have your media relations, external affairs or lawyer with you, because some questions do not legally have to be answered. For example, if the reporter asks something that deals with confidential or proprietary information, say that the question can't be answered for that reason.

Environmental Groups

It is important to consider the feelings and goals of environmental groups such as the Sierra Club, Friends Of The Earth and others. Their involvement in significant environmental issues has had a major impact in the past. The old belief that environmental groups are 100% against industry and industry 100% against the environment is changing for the better. However, both of these components of society need to continue working on bringing their "perceived" polarized purposes closer to a middle ground.

Public/Customers/Community

The public in general fears and misunderstands HMW. The NIMBY feelings sum up the attitude on more than just disposal. It should be kept in mind that the transportation, storage and use of HMW are also mistrusted. Public awareness meetings and communications (fact sheets) are often necessary to keep the public informed. Environmental awareness conferences are another good way to promote positive public attitude and understanding. These events can be hosted by anyone, including the hazardous waste generators who are operating in the community.

Organized Crime

Organized crime has been and still is deeply involved in HW. Investigations have shown that mafia connections to some transporters, disposal sites, politicians and agency individuals still exist. If an HMW manager keeps this in mind he/she will remember to question HW services offered below market cost and will stay current on HW vendor audits.

The past and present impacts of organized crime on the history of hazardous waste management have been severe. Organized crime has had as significant an influence on the history of hazardous waste management as have the agencies and environmental groups. Of course, this influence has largely been negative. This should dramatically illustrate one of many reasons for close and careful hazardous material and waste management, especially when you consider the fact that organized crime is still present today.

Steven J. Madonna, Deputy Attorney General in charge of New Jersey's Environment Prosecution Section, summed it up very clearly in a statement made to the Dingell Committee Hearings: "Mr. Chairman, I believe that the waste industry is probably one of the most violent industries that exists in the country today. There have been murders, threats, assaults, arson, intimidations. It seems to be unfortunately a way of life in that industry. I am not talking about the day-to-day collector, but the type of controls imposed are such that in fact organized crime plays a significant role in that industry" [Block 1985].

Authors Alan Block and Frank Scarpitti drive the point home further in their book *Poisoning For Profit*. They state, "In the case of toxic waste disposal, organized crime has become so entrenched in the industry that syndicates now control not only the dumps, the unions and the trade associations, but also the politicians and environmental officials responsible for enforcing disposal laws and regulations" [Block 1985].

Bounty Hunters

Bounty hunters are a new breed of individual that the HMW professional must consider. Spawned out of certain regulations such as California Prop. 65, bounty hunters can obtain 10% of a Prop. 65 violation settlement in some situations. Unfortunately, the motive of this individual may be more profit-oriented than environmental. HMW plans must be comprehensive to protect against bounty hunter attack.

Agencies

The primary goal of most HMW agencies is to minimize environmental and employee impacts. In order to do this, they must establish and enforce regulations, advise and provide various other services. When considering the agency part of the equation, it is good to keep in mind that they have internal problems just like any other organization. For example, there is conflict between some agencies, overlap in responsibilities, and most have budget problems.

Many agency individuals are not rewarded for taking risks and are encouraged to make very conservative decisions. This approach is sometimes followed in order to minimize court involvement. If this is kept in mind during communications with agencies, a more meaningful relationship will be established. Early, frequent and open discussions are of extreme value to all involved.

The environmental professional should remember that the organization's relationship with agencies is long-term and that the agency is staffed with human beings. These agency individuals rightly take their jobs seriously and resent being embarrassed or misled. Short-term victories (permit granting, etc.), won through unrealistic promises or less than full disclosure, are sure ways of spoiling the organization's credibility for a long period of time. Communication to the agency needs to be prompt and factual [Meehan 1993].

Politicians

Since politicians are the primary authors of many of the new HMW bills and acts, they must be considered in the management plan also. In addition to wanting to help minimize environmental impacts, some politicians must also worry about being reelected. An understanding of their perspective will help the HMW professional better anticipate and deal with the legislation they create.

Industry

The basic goal of industry is to make a profit and stay in business. To stay in business, most industries are recognizing that they must comply with regulations, minimize impacts and reduce hazardous materials purchased. Proper HMW handling, in all parts of the world, makes good business sense. More and more each day, investors are demanding that industry does its part in protecting the environment.

Almost any action that an organization takes to minimize an HMW impact may affect other industries. For example, if a hazardous material is phased out of a process, the manufacturer of that material may protest, especially if it is a significant loss in business. By anticipating possible reactions such as this, an organization can be better prepared to handle them when and if they occur.

Consultants

HMW consultants are also an important part of the entire system. As with industry, they need to make a profit to stay in business. To do this, they must be competitive and become experts in many HMW topics. Often, consultants become an interface between agencies and industry, implement programs and do research.

Academic Research

The primary goal of individuals in HMW academic research is to obtain a better understanding of environmental impacts. As with all of the other organizations mentioned, they must also worry about budgets. Their perspective must be considered, since they are an important component of the HMW system.

Case Study

ICMESA—Seveso, Italy

This study illustrates how human indecision and social and political pressure can endanger lives and health. On July 10, 1976, one of the most devastating accidents in the recent history of hazardous waste management occurred in the town of Seveso in Northern Italy. The incident resulted in needed regulations governing hazardous waste. Following the incident, the European Economic Community set up a waste control program.

The Industrie Chemiche Meda Societa Aromia (ICMESA), which is a division of Hoffman-La Roche, manufactures disinfectants, deodorants, cosmetics and herbicides; 2,4,5-trichlorophenol is manufactured as an intermediate product. An explosion took place at the plant when shift workers allowed a reactor processing trichlorophenol to overheat. This resulted in blowing out a rupture disk in a safety valve. The ensuing chemical reaction produced 2,3,7,7-tetrachlorodi-benzo-p-dioxin (TCDD), which was released to the atmosphere. Approximately one kilogram of dioxin was discharged into the air above the plant.

The dioxin cloud travelled several hundred acres toward Milan. The vicinity of the plant was highly contaminated. Neighboring areas were contaminated to a lesser degree. Hundreds of birds, cats, chickens and other animals died. Liver damage was found in autopsies.

Immediately following the incident, the company responded that not much had escaped from the facility. Within two weeks, however, trees started to lose their leaves, and dogs, birds and rabbits were dying. Children broke out with chloracne and some people reported having dangerous side effects. On July 16, workers refused to go back to work. Approximately two weeks after the incident, the company admitted that there was contamination. During the period of indecision, however, people continued to live in highly dangerous conditions while the contamination migrated in surface and ground water.

Three contamination zones were established. Zone A covered about 285 acres and contained the highest concentration of dioxin. Within two weeks, approximately 730 residents were evacuated from this zone. Zone B covered 500 acres, and the residents were warned not to eat the fruits and vegetables. Zone C covered 3500 acres, and approximately 4280 residents were warned not to eat their crops. Local waterways and property suffered severe contamination.

A plan was developed to decontaminate the 12 factory buildings and 100 houses. Incineration and on-site burial were agreed upon. Some 500,000 tons of soil, vegetation, building materials and cleanup tools were generated.

The accident also had sociopolitical repercussions. Pregnant women were told that they would bear deformed babies. Fear of malformed fetuses prompted doctors in Seveso to warn their patients to have abortions. The warning touched off a debate in Catholic Italy, where Church authorities could not condone abortion under any circumstance and stated that the Church would find other people to take care of deformed children.

After one year, over 81,000 domestic animals were dead from the dioxin poisoning. Only one pregnant cow of 13 gave birth to a healthy calf. The remaining had spontaneous abortions or deformed calves. Approximately 280 children developed chloracne, blurred vision, liver damage and depression. Among the women of child-bearing age, there was a one-to-four rate of spontaneous abortions. These women were also found to have incurred damage to their chromosomes and those of their fetuses.

In June 1982, 41 drums were shipped off-site for disposal. The drums were removed from the site and shipped to an unspecified location. The wastes were not labeled properly, and the point of origin was marked incorrectly. After nine months of searching for the missing drums, they turned up in a French slaughterhouse. The waste was finally shipped to Switzerland for incineration, almost two and one half years after leaving the site.

In April 1983, five ICMESA employees were tried in an Italian criminal court for the incident at Seveso. The five defendants were convicted of criminal negligence. The costs of cleanup have amounted to $200 million [Wentz 1989].

CHAPTER

SEVENTEEN

Conclusion

By concluding this book with a discussion emphasizing the human element does not imply that the non-human species are less important. In fact, all of the other chapters present regulations, assessments and controls that help to protect the entire environment.

A brief sampling of many HMW topics has been presented in this book. Because of the great number of topics in this field, only the essence of each was provided. A special attempt was made to present only essential information that will help the reader achieve and exceed regulatory compliance.

Organizations that expect to remain viable today and into the future must exceed what the regulations specify. Doing this in a cost-effective way is the real challenge, but it must be done. Customers all around the world are now starting to demand products and services from proactive, environmentally conscious organizations. By around 1996, it will be easy for these customers to select the environmentally conscious suppliers. All they will have to do is look for organizations that have been certified under the new ISO standards that apply to environmental, health and safety. Easy for the customer, but extremely hard for the supplier. This book has presented a large portion of what it will take to get the certification.

Hopefully, the reader sensed throughout the book and especially in Chapter 16, "The Human Element" that there are many different players in the HMW game, each with their own perspective. Application of this concept in real life will make an HMW career more meaningful and productive. It will also help the reader better serve the goals of as many important stakeholders as possible, including the plants and animals of the entire ecosystem.

Another common theme in some of the chapters was the life cycle of an HMW. Many of the chapters were presented in the life cycle order (i.e., use and storage, shipment, treatment and disposal). By using the life cycle as an organizing tool, the HMW professional will have a meaningful framework for placement of all the details they encounter later.

The hazardous material and waste management field is extremely broad and complex. Individuals in this field must be familiar with many different disciplines such as toxicology, ecology, engineering, chemistry, biology, industrial hygiene and many others. This document has attempted to present just a sampling or overview of some of these areas that have an important impact on hazardous material and waste management.

If we all work together to manage and minimize the hazardous materials and wastes that are an intimate part of our lives, every component of the ecosystem will benefit. Is there a more worthwhile project?

A

List of Hazardous Substances and Reportable Quantities

Appendix A is an example of a list of hazardous materials and wastes. This particular list is fron 40 CFR Chapter 1 and was prepared by the U.S. Environmental Protection Agency (EPA). The EPA has other lists, as does the U.S. Department of Transportation and numerous other federal, state and local agencies. None of these lists are all inclusive; therefore, it is necessary to check several, depending on the specific situation.

There are detailed notes at the end of Appendix A, which will aid the reader in understanding the information presented. Not shown, however, is the following additional information about column headings:

- Hazardous Substance Column—This refers to the scientific name of the chemical or hazardous waste that is regulated by the EPA.

- CASRN Column—The CASRN is a Chemical Abstract Service number. There is a separate CAS number for every chemical. Complex mixtures might have several CAS numbers. Most hazardous wastes are mixtures and would be composed of various used chemicals, each with its own CAS number.

- Regulatory Synonyms Column—If a chemical or hazardous waste is commonly known by a different name, it is shown in this column.

- RQ Column—If this amount of the chemical or hazardous waste is spilled, it must be reported to a regulatory agency.

Hazardous substance	CASRN	Statutory RQ	Code†	RCRA waste Number	Final RQ Category	Pounds (Kg)
Acenaphthene	83329	1*	2		B	100 (45.4)
Acenaphthylene	208968	1*	2		D	5000 (2270)
Acetaldehyde Ethanal	75070	1000	1,4	U001	C	1000 (454)
Acetaldehyde, chloro- Chloroacetaldehyde	107200	1*	4	P023	D	1000 (454)
Acetaldehyde, trichloro- Chloral	75876	1*	4	U034	D	5000 (2270)
Acetamide, N-(aminothioxomethyl)- 1-Acetyl-2-thiourea	591082	1*	4	P002	C	1000 (454)
Acetamide, N-(4-ethoxyphenyl)- Phenacetin	62442	1*	4	U187	B	100 (45.4)
Acetamide, 2-fluoro- Fluoroacetamide	640197	1*	4	P057	B	100 (45.4)
Acetamide, N-9H-fluoren-2-yl- 2-Acetylaminofluorene	53963	1*	1	U005	X	1 (0.454)
Acetic acid	64197	1000			D	5000 (2270)
Acetic acid (2,4-dichlorophenoxy)- 2,4-D Acid 2,4-D, salts and esters	94757	100	1,4	U240	B	100 (45.4)
Acetic acid, Lead(2+) salt Lead acetate	301042	5000	1,4	U144	A	10 (4.54)
Acetic acid, thallium (1+) salt Thallium(I) acetate	563688	1*	1,4	U214	B	100 (45.4)
Acetic acid, (2,4,5-trichlorophenoxy)- 2,4,5-T 2,4,5-T acid	93765	100	1,4	U232	C	1000 (454)
Acetic acid, ethyl ester Ethyl acetate	141786	1*	4	U112	D	5000 (2270)
Acetic acid, fluoro-, sodium salt Fluoroacetic acid, sodium salt	62748	1*	4	P058	A	10 (4.54)
Acetic anhydride	108247	1000	1		D	5000 (2270)
Acetone 2-Propanone	67641	1*	4	U002	D	5000 (2270)
Acetone cyanohydrin Propanenitrile, 2-hydroxy-2-methyl-2- Methyllactonitrile.	75865	10	1,4	P069	A	10 (4.54)
Acetonitrile	75058	1*	4	U003	D	5000 (2270)
Acetophenone Ethanone, 1-phenyl-	98862	1*	4	U004	D	5000 (2270)
2-Acetylaminofluorene Acetamide, N-9H-fluoren-2-yl-	53963	1*	4	U005	X	1 (0.454)
Acetyl bromide	506967	5000	1		D	5000 (2270)
Acetyl chloride	75365	5000	1,4	U006	D	5000 (2270)
1-Acetyl-2-thiourea Acetamide, N-(aminothioxomethyl)-	591082	1*	4	P002	C	1000 (454)
Acrolein 2-Propenal	107028	1	1,2,4	P003	X	1 (0.454)
Acrylamide 2-Propenamide	79061	1*	4	U007	D	5000 (2270)
Acrylic acid 2-Propenoic acid	79107	1*	4	U008	D	5000 (2270)
Acrylonitrile 2-Propenenitrile	107131	100	1,2,4	U009	B	100 (45.4)
Adipic acid	124049	5000	1		D	5000 (2270)
Aldicarb Propanal, 2-methyl-2-(methylthio)-, O-[(methylamino)carbonyl]oxime.	116063	1*	4	P070	X	1 (0.454)
Aldrin 1,4,5,8-Dimethanonaphthalene, 1,2,3,4,10,10-hexachloro-1,4,4a,5,8,8a-hexahydro-, (1alpha, 4alpha,4abeta,5alpha,8alpha,8abeta)-.	309002	1	1,2,4	P004	X	1 (0.454)
Allyl alcohol 2-Propen-1-ol	107186	100	1,4	P005	B	100 (45.4)
Allyl chloride	107051	1000	1		C	1000 (454)
Aluminum phosphide	20859738	1*	4	P006	B	100 (45.4)

Substance	Synonym	CAS No.	RQ	Category	Waste No.	Code	Final RQ lb (kg)
Aluminum sulfate		10043013	5000	1		D	5000 (2270)
5-(Aminomethyl)-3-isoxazolol	Muscimol 3(2H)-Isoxazolone, 5-(aminomethyl)-	2763964	1*	4	P007	C	1000 (454)
4-Aminopyridine	4-Pyridinamine	504245	1*	4	P008	C	1000 (454)
Amitrole	1H-1,2,4-Triazol-3-amine	61825	1*	4	U011	A	10 (4.54)
Ammonia		764417	100	1		B	100 (45.4)
Ammonium acetate		631618	5000	1		D	5000 (2270)
Ammonium benzoate		1863634	5000	1		D	5000 (2270)
Ammonium bicarbonate		1066337	5000	1		D	5000 (2270)
Ammonium bichromate		7789095	1000	1		A	10 (4.54)
Ammonium bifluoride		1341497	5000	1		B	100 (45.4)
Ammonium bisulfite		10192300	5000	1		D	5000 (2270)
Ammonium carbamate		1111780	5000	1		D	5000 (2270)
Ammonium carbonate		506876	5000	1		D	5000 (2270)
Ammonium chloride		12125029	5000	1		D	5000 (2270)
Ammonium chromate		7788989	1000	1		A	10 (4.54)
Ammonium citrate, dibasic		3012655	5000	1		D	5000 (2270)
Ammonium fluoborate		13826830	5000	1		D	5000 (2270)
Ammonium fluoride		12125018	5000	1		B	100 (45.4)
Ammonium hydroxide		1336216	1000	1		C	1000 (454)
Ammonium oxalate		6009707 / 5972736 / 14258492	5000	1		D	5000 (2270)
Ammonium picrate	Phenol, 2,4,6-trinitro-, ammonium salt	131748	1*	4	P009	A	10 (4.54)
Ammonium silicofluoride		16919190	1000	1		C	1000 (454)
Ammonium sulfamate		7773060	5000	1		D	5000 (2270)
Ammonium sulfide		12135761	5000	1		B	100 (45.4)
Ammonium sulfite		10196040	5000	1		D	5000 (2270)
Ammonium tartrate		14307438 / 3164292	5000	1		D	5000 (2270)
Ammonium thiocyanate		1762954	5000	1		D	5000 (2270)
Ammonium vanadate	Vanadic acid, ammonium salt	7803556	1*	4	P119	C	1000 (454)
Amyl acetate		628637	1000	1		D	5000 (2270)
iso-Amyl acetate		123922		1		D	5000 (2270)
sec-Amyl acetate		626380		1		D	5000 (2270)
tert-Amyl acetate		625161		1		D	5000 (2270)
Aniline	Benzenamine	62533	1000	1,4	U012	C	1000 (454)
Anthracene		120127	1*	2		B	100 (45.4)
Antimony††		7440360	1*	2		C	1000 (454)
ANTIMONY AND COMPOUNDS		N.A.		2		C	1000 (454)
Antimony pentachloride		7647189	1000	1		C	1000 (454)
Antimony potassium tartrate		28300745	1000	1		C	1000 (454)
Antimony tribromide		7789619	1000	1		C	1000 (454)
Antimony trichloride		10025919	1000	1		C	1000 (454)
Antimony trifluoride		7783564	1000	1		C	1000 (454)
Antimony trioxide		1309644	1000	1		C	1000 (454)
Argentate(1-), bis(cyano-C)-, potassium	Potassium silver cyanide	506616	1*	4	P099	X	1 (0.454)
Aroclor 1016	POLYCHLORINATED BIPHENYLS (PCBs)	12674112	10	1,2		X	1 (0.454)
Aroclor 1221	POLYCHLORINATED BIPHENYLS (PCBs)	11104282	10	1,2		X	1 (0.454)
Aroclor 1232	POLYCHLORINATED BIPHENYLS (PCBs)	11141165	10	1,2		X	1 (0.454)
Aroclor 1242	POLYCHLORINATED BIPHENYLS (PCBs)	53469219	10	1,2		X	1 (0.454)

Hazardous substance	CASRN	Regulatory synonyms	Statutory RQ	Statutory Code†	Statutory RCRA waste Number	Final RQ Category	Final RQ Pounds (Kg)
Aroclor 1248	12672296	POLYCHLORINATED BIPHENYLS (PCBs)	10	1,2		X	1 (0.454)
Aroclor 1254	11097691	POLYCHLORINATED BIPHENYLS (PCBs)	10	1,2		X	1 (0.454)
Aroclor 1260	11096825	POLYCHLORINATED BIPHENYLS (PCBs)	10	1,2		X	1 (0.454)
Arsenic ††	7440382		1*	2,3		X	1 (0.454)
Arsenic acid	1327522	Arsenic acid H3AsO4	1*	4	P010	X	1 (0.454)
Arsenic acid H3AsO4	7778394	Arsenic acid	1*	4	P010	X	1 (0.454) **
ARSENIC AND COMPOUNDS	N.A.			2		X	
Arsenic disulfide	1303328		5000	1		X	1 (0.454)
Arsenic oxide As2O3	1327533	Arsenic trioxide	5000	1,4	P012	X	1 (0.454)
Arsenic pentoxide	1303282	Arsenic oxide As2O5	5000	1,4	P011	X	1 (0.454)
Arsenic oxide As2O5	1303282	Arsenic pentoxide	5000	1,4	P011	X	1 (0.454)
Arsenic trichloride	7784341		5000	1		X	1 (0.454)
Arsenic trioxide	1327533	Arsenic oxide As2O3	5000	1,4	P012	X	1 (0.454)
Arsenic trisulfide	1303339		5000	1		X	1 (0.454)
Arsine, diethyl-	692422	Diethylarsine	1*	4	P038	X	1 (0.454)
Arsinic acid, dimethyl-	75605	Cacodylic acid	1*	4	U136	X	1 (0.454)
Arsonous dichloride, phenyl-	696286	Dichlorophenylarsine	1*	4	P036	X	1 (0.454)
Asbestos †††	1332214		1*	2,3		B	100 (45.4)
Auramine	492808	Benzenamine, 4,4'-carbonimidoylbis (N,N-dimethyl-	1*	4	U014	X	1 (0.454)
Azaserine	115026	L-Serine, diazoacetate (ester)	1*	4	U015	X	1 (0.454)
Aziridine	151564	Ethylenimine	1*	4	P054	X	1 (0.454)
Aziridine, 2-methyl-	75558	1,2-Propylenimine	1*	4	P067	X	1 (0.454)
Azirino[2',3':3,4]pyrrolo[1,2-a]indole-4,7-dione,6-amino-8-[[(aminocarbonyl)oxy]methyl]-1,1a,2,8,8a,8b-hexahydro-8a-methoxy-5-methyl-,[1aS-(1aalpha,8beta,8aalpha,8balpha)]-	50077	Mitomycin C	1*	4	U010	A	10 (4.54)
Barium cyanide	542621		10	1,4	P013	A	10 (4.54)
Benz[j]aceanthrylene, 1,2-dihydro-3-methyl-	56495	3-Methylcholanthrene	1*	4	U157	A	10 (4.54)
Benz[c]acridine	225514		1*	4	U016	B	100 (45.4)
Benzal chloride	98873	Benzene, dichloromethyl-	1*	4	U017	D	5000 (2270)
Benzamide, 3,5-dichloro-N-(1,1-dimethyl-2-propynyl)-	23950585	Pronamide	1*	4	U192	D	5000 (2270)
Benz[a]anthracene	56553	Benzo[a]anthracene	1*	2,4	U018	A	10 (4.54)
1,2-Benzanthracene	56553	Benz[a]anthracene	1*	2,4	U018	A	10 (4.54)
Benz[a]anthracene, 7,12-dimethyl-	57976	7,12-Dimethylbenz[a]anthracene	1*	4	U094	X	1 (0.454)
Benzenamine	62533	Aniline	1000	4	U012	D	5000 (2270)
Benzenamine, 4,4'-carbonimidoylbis (N,N-dimethyl-	492808	Auramine	1*	1,4	U014	B	100 (45.4)
Benzenamine, 4-chloro-	106478	p-Chloroaniline	1*	4	P024	C	1000 (454)

CAS Number	Hazardous Substance and Synonyms	Statutory RQ	Statutory Code	RCRA Waste No.	Category	Final RQ Pounds (Kg)
N.A.	**BERYLLIUM AND COMPOUNDS**					**
7787475	Beryllium chloride	5000	2		X	1 (0.454)
7440417	Beryllium††	1*	1		A	10 (4.54)
7787497	Beryllium dust††	5000	2,3,4	P015	X	1 (0.454)
13597994	Beryllium fluoride	5000	1		X	1 (0.454)
7787555	Beryllium nitrate	5000	1		X	1 (0.454)
319846	alpha—BHC	1*	2		A	10 (4.54)
319857	beta—BHC	1*	2		X	1 (0.454)
319868	delta—BHC	1*	2		X	1 (0.454)
58899	gamma—BHC Cyclohexane, 1,2,3,4,5,6-hexachloro-(1alpha,2alpha,3beta,4alpha,5alpha,6beta)- Hexachlorocyclohexane (gamma isomer) Lindane.	1	1,2,4	U129	X	1 (0.454)
1464535	2,2'-Bioxirane	1*	4	U085	A	10 (4.54)
92875	Benzidine [1,1'-Biphenyl]-4,4'diamine	1*	2,4	U021	X	1 (0.454)
91941	3,3'-Dichlorobenzidine [1,1'-Biphenyl]-4,4'diamine,3,3'dichloro-	1*	2,4	U073	X	1 (0.454)
119904	3,3'-Dimethoxybenzidine [1,1'-Biphenyl]-4,4'-diamine,3,3'dimethoxy-	1*	2,4	U091	B	100 (45.4)
119937	3,3'-Dimethylbenzidine [1,1'Biphenyl]-4,4'-diamine,3,3'-dimethyl-	1*	4	U095	A	10 (4.54)
111444	Dichloroethyl ether Ethane,1,1'-oxybis[2-chloro-	1*	2,4	U025	A	10 (4.54)
111911	Bis(2-chloroethoxy) methane Dichloromethoxy ethane	1*	2,4	U024	C	1000 (454)
117817	Bis (2-ethylhexyl)phthalate Ethane, 1,1'-[methylenebis(oxy)]bis[2-chloro- 1,2-Benzenedicarboxylic acid, [bis(2-ethylhexyl)] ester	1*	2,4	U028	B	100 (45.4)
598312	Bromoacetone 2-Propanone, 1-bromo-	1*	4	P017	C	1000 (454)
75252	Bromoform Methane, tribromo-	1*	2,4	U225	B	100 (45.4)
101553	4-Bromophenyl phenyl ether Benzene, 1-bromo-4-phenoxy-	1*	2,4	U030	B	100 (45.4)
357573	Brucine Strychnidin-10-one, 2,3-dimethoxy-	1*	2,4	P018	B	100 (45.4)
87683	Hexachlorobutadiene 1,3-Butadiene, 1,1,2,3,4,4-hexachloro-	1*	2,4	U128	X	1 (0.454)
924163	N-Nitrosodi-n-butylamine 1-Butanamine, N-butyl-N-nitroso-	1*	4	U172	A	10 (4.54)
71363	n-Butyl alcohol 1-Butanol	1*	4	U031	D	5000 (2270)
78933	Methyl ethyl ketone (MEK) 2-Butanone	1*	4	U159	D	5000 (2270)
1338234	Methyl ethyl ketone peroxide 2-Butanone peroxide	1*	4	U160	A	10 (4.54)
39196184	Thiofanox 2-Butanone, 3,3-dimethyl-1-(methylthio)-, O[(methylamino)carbonyl] oxime.	1*	4	P045	B	100 (45.4)
123739 4170303	Crotonaldehyde 2-Butenal	100	1,4	U053	B	100 (45.4)
764410	1,4-Dichloro-2-butene 2-Butene, 1,4-dichloro-	1*	4	U074	X	1 (0.454)
303344	Lasiocarpine 2-Butenoic acid, 2-methyl-, 7[[2,3-dihydroxy-2-(1-methoxyethyl)-3-methyl-1-oxobutoxy)methyl]-2,3,5,7a-tetrahydro-1H-pyrrolizin-1-yl ester, [1S-[1alpha(Z),7(2S',3R'),7aalpha]]-	1*	4	U143	A	10 (4.54)
123864	Butyl acetate	5000	1		D	5000 (2270)
110190	iso-Butyl acetate					
105464	sec-Butyl acetate					
540885	tert-Butyl acetate					
71363	n-Butyl alcohol 1-Butanol	1*	4	U031	D	5000 (2270)
109739	Butylamine	1000	1		C	1000 (454)
78819	iso-Butylamine					

Hazardous substance	CASRN	Regulatory synonyms	Statutory RQ	Statutory Code†	Statutory RCRA waste Number	Final RQ Category	Final RQ Pounds (Kg)
sec-Butylamine	513495		1*	2		B	100 (45.4)
tert-Butylamine	13952846		100	1,2,4		A	10 (4.54)
Butyl benzyl phthalate	75649						
n-Butyl phthalate	85687	Di-n-butyl phthalate; Dibutyl phthalate; 1,2-Benzenedicarboxylic acid, dibutyl ester		1	U069	D	5000 (2270)
	84742						
Butyric acid	107926		5000	1		X	1 (0.454)
iso-Butyric acid	79312		1*	4		A	10 (4.54)
Cacodylic acid	75605	Arsinic acid, dimethyl-	1*	2	U136	A	10 (4.54)**
Cadmium ††	7440439		100	1		A	10 (4.54)
Cadmium acetate	543908		1*	2		A	10 (4.54)
CADMIUM AND COMPOUNDS	N.A.						
Cadmium bromide	7789426		100	1		X	1 (0.454)
Cadmium chloride	10108642		100	1		X	1 (0.454)
Calcium arsenate	7778441		1000	1		A	10 (4.54)
Calcium arsenite	52740166		1000	1		A	10 (4.54)
Calcium carbide	75207		5000	1		A	10 (4.54)
Calcium chromate	13765190	Chromic acid H2CrO4, calcium salt	1000	1,4	U032	A	10 (4.54)
Calcium cyanide	592018	Calcium cyanide Ca(CN)2	10	1,4	P021	C	1000 (454)
Calcium cyanide Ca(CN)2	592018	Calcium cyanide	10	1,4	P021	A	10 (4.54)
Calcium dodecylbenzenesulfonate	26264062		1000	1		X	1 (0.454)
Calcium hypochlorite	7778543		100	1		A	10 (4.54)
Camphene, octachloro-	8001352	Toxaphene	1	1,2,4	P123	B	100 (45.4)
Captan	133062		10	1		X	1 (0.454)
Carbamic acid, ethyl ester	51796	Ethyl carbamate (urethane)	1*	4	U238	X	1 (0.454)
Carbamic acid, methylnitroso-, ethyl ester	615532	N-Nitroso-N-methylurethane	1*	4	U178	D	5000 (2270)
Carbamic chloride, dimethyl-	79447	Dimethylcarbamoyl chloride	1*	4	U097	B	100 (45.4)
Carbamodithioic acid, 1,2-ethanediylbis, salts & esters	111546	Ethylenebisdithiocarbamic acid, salts & esters	1*	4	U114	B	100 (45.4)
Carbamothioic acid, bis(1-methylethyl)-, S-(2,3-dichloro-2-propenyl) ester	2303164	Diallate	1*	4	U062	A	10 (4.54)
Carbaryl	63252		100	1		B	100 (45.4)
Carbofuran	1563662		10	1		C	1000 (454)
Carbon disulfide	75150		5000	1,4	P022	B	100 (45.4)
Carbon oxyfluoride	353504	Carbonic difluoride	1*	4	U033	C	1000 (454)
Carbon tetrachloride	56235	Methane, tetrachloro-	5000	1,2,4	U211	A	10 (4.54)
Carbonic acid, dithallium(1+) salt	6533739	Thallium(I) carbonate	1*	4	U215	B	100 (45.4)
Carbonic dichloride	75445	Phosgene	5000	1,4	P095	A	10 (4.54)
Carbonic difluoride	353504	Carbon oxyfluoride	1*	4	U033	C	1000 (454)

Substance	Chemical name	CAS No.			RCRA No.		RQ lb (kg)
Carbonochloridic acid, methyl ester	Methyl chlorocarbonate; Methyl chloroformate	79221	1*	4	U156	C	1000 (454)
Chloral	Acetaldehyde, trichloro-	75876	1*	4	U034	D	5000 (2270)
Chlorambucil	Benzenebutanoic acid, 4-[bis(2-chloroethyl)amino]-	305033	1*	4	U035	A	10 (4.54)
Chlordane	Chlordane, alpha & gamma isomers; Chlordane, technical; 4,7-Methano-1H-indene, 1,2,4,5,6,7,8,8-octachloro-2,3,3a,4,7,7a-hexahydro-	57749	1	1,2,4	U036	X	1 (0.454)
CHLORDANE (TECHNICAL MIXTURE AND METABOLITES)		N.A.	1*	2			**
Chlordane, alpha & gamma isomers	Chlordane; Chlordane, technical; 4,7-Methano-1H-indene, 1,2,4,5,6,7,8,8-octachloro-	57749	1	1,2,4	U036	X	1 (0.454)
Chlordane, technical	Chlordane, alpha & gamma isomers; Chlordane, technical; 4,7-Methano-1H-indene, 1,2,4,5,6,7,8,8-octachloro-2,3,3a,4,7,7a-hexahydro-	57749	1	1,2,4	U036	X	1 (0.454)
CHLORINATED BENZENES		N.A.	1*	2			**
CHLORINATED ETHANES		N.A.	1*	2			**
CHLORINATED NAPHTHALENE		N.A.	1*	2			**
CHLORINATED PHENOLS		N.A.	1*	1			**
Chlorine		7782505	10	4		A	10 (4.54)
Chlornaphazine	Naphthalenamine, N,N'-bis(2-chloroethyl)-	494031	1*	4	U026	B	100 (45.4)
Chloroacetaldehyde	Acetaldehyde, chloro-	107200	1*	2	P023	C	1000 (454)
CHLOROALKYL ETHERS		N.A.	1*	1,2,4			**
p-Chloroaniline	Benzenamine, 4-chloro-	106478	100	4	P024	C	1000 (454)
Chlorobenzene	Benzene, chloro-	108907	1*	2	U037	B	100 (45.4)
Chlorobenzilate	Benzeneacetic acid, 4-chloro-alpha-(4-chlorophenyl)-alpha-hydroxy-, ethyl ester.	510156	1*	4	U038	A	10 (4.54)
4-Chloro-m-cresol	p-Chloro-m-cresol; Phenol, 4-chloro-3-methyl-	59507	1*	2,4	U039	D	5000 (2270)
p-Chloro-m-cresol	Phenol, 4-chloro-3-methyl-; 4-Chloro-m-cresol	59507	1*	2,4	U039	D	5000 (2270)
Chlorodibromomethane		124481	1*	2		B	100 (45.4)
Chloroethane		75003	1*	2		B	100 (45.4)
2-Chloroethyl vinyl ether	Ethene, 2-chloroethoxy-	110758	1*	2,4	U042	C	1000 (454)
Chloroform	Methane, trichloro-	67663	5000	1,2,4	U044	A	10 (4.54)
Chloromethyl methyl ether	Methane, chloromethoxy-	107302	1*	4	U046	A	10 (4.54)
beta-Chloronaphthalene	Naphthalene, 2-chloro-; 2-Chloronaphthalene	91587	1*	2,4	U047	D	5000 (2270)
2-Chloronaphthalene	beta-Chloronaphthalene; Naphthalene, 2-chloro-	91587	1*	2,4	U047	D	5000 (2270)
2-Chlorophenol	o-Chlorophenol; Phenol, 2-chloro-	95578	1*	2,4	U048	B	100 (45.4)
o-Chlorophenol	Phenol, 2-chloro-; 2-Chlorophenol	95578	1*	2,4	U048	B	100 (45.4)
4-Chlorophenyl phenyl ether		7005723	1*	2		D	5000 (2270)
1-(o-Chlorophenyl)thiourea	Thiourea, (2-chlorophenyl)-	5344821	1*	4	P026	B	100 (45.4)
3-Chloropropionitrile	Propanenitrile, 3-chloro-	542767	1*	4	P027	C	1000 (454)

Hazardous substance	CASRN	Regulatory synonyms	RQ	Code†	RCRA waste Number	Category	Pounds (Kg)
Chlorosulfonic acid	7790945		1000	1		C	1000 (454)
4-Chloro-o-toluidine, hydrochloride	3165933	Benzenamine, 4-chloro-2-methyl-, hydrochloride	1*	4	U049	B	100 (45.4)
Chlorpyrifos	2921882		1	1		X	1 (0.454)
Chromic acetate	1066304		1000	1		C	1000 (454)
Chromic acid	11115745 7738945		1000	1		A	10 (4.54)
Chromic acid H2CrO4, calcium salt	13765190	Calcium chromate	1000	1,4	U032	A	10 (4.54)
Chromic sulfate	10101538		1000	2		C	1000 (454)
Chromium††	7440473		1*	2		D	5000 (2270) **
CHROMIUM AND COMPOUNDS	N.A.						
Chromous chloride	10049055		1000	1		C	1000 (454)
Chrysene	218019	1,2-Benzphenanthrene	1*	2,4	U050	B	100 (45.4)
Cobaltous bromide	7789437		1000	1		C	1000 (454)
Cobaltous formate	544183		1000	1		C	1000 (454)
Cobaltous sulfamate	14017415		1*	1		C	1000 (454)
Coke Oven Emissions	N.A.		1*	3		X	1 (0.454)
Copper cyanide	544923	Copper cyanide CuCN	1*	4	P029	A	10 (4.54)
Copper††	7440508		1*	2		D	5000 (2270) **
COPPER AND COMPOUNDS	N.A.						
Copper cyanide	544923	Copper cyanide CuCN	1*	4	P029	A	10 (4.54)
Coumaphos	56724		10	1		A	10 (4.54)
Creosote	8001589		1*	4	U051	X	1 (0.454)
Cresol(s)	1319773	Cresylic acid	1000	1,4	U052	C	1000 (454)
m-Cresol	108394	Phenol, methyl-					
o-Cresol	95487	m-Cresylic acid					
p-Cresol	106445	o-Cresylic acid					
Cresylic acid	1319773	p-Cresylic acid	1000	1,4	U052	C	1000 (454)
m-Cresol	108394	Phenol, methyl-					
o-Cresol	95487	m-Cresylic acid					
p-Cresol	106445	o-Cresylic acid					
Crotonaldehyde	123739 4170303	2-Butenal	100	1,4	U053	B	100 (45.4)
Cumene	98828	Benzene, 1-methylethyl-	1*	4	U055	D	5000 (2270)
Cupric acetate	142712		100	1		B	100 (45.4)
Cupric acetoarsenite	12002038		100	1		X	1 (0.454)
Cupric chloride	7447394		10	1		A	10 (4.54)
Cupric nitrate	3251238		100	1		B	100 (45.4)
Cupric oxalate	5893663		10	1		B	100 (45.4)
Cupric sulfate	7758987		10	1		A	10 (4.54)

Substance	CAS Number	Description / Synonyms	Statutory RQ	Code	RCRA Waste No.	Category	Final RQ
Cupric sulfate, ammoniated	10380297		100	1		B	100 (45.4)
Cupric tartrate	815827		100	1		B	100 (45.4)**
CYANIDES	N.A.						
Cyanides (soluble salts and complexes) not other-wise specified	57125		1*	2	P030	A	10 (4.54)
Cyanogen	460195	Ethanedinitrile	1*	4	P031	B	100 (45.4)
Cyanogen bromide (CN)Br	506683	Cyanogen bromide (CN)Br	1*	4	U246	C	1000 (454)
Cyanogen bromide	506683	Cyanogen bromide	1*	4	U246	C	1000 (454)
Cyanogen chloride (CN)Cl	506774	Cyanogen chloride (CN)Cl	10	1,4	P033	A	10 (4.54)
Cyanogen chloride	506774	Cyanogen chloride (CN)Cl	10	1,4	P033	A	10 (4.54)
2,5-Cyclohexadiene-1,4-dione	106514	p-Benzoquinone	1*	1,4	U197	A	10 (4.54)
Cyclohexane	110827	Benzene, hexahydro-	1000	1,4	U056	C	1000 (454)
Cyclohexane, 1,2,3,4,5,6-hexachloro-,(1alpha,2alpha,3beta,4alpha,5alpha, 6beta)-	58899	gamma—BHC; Hexachlorocyclohexane (gamma isomer)[Lindane; Lindane	1	1,2,4	U129	X	1 (0.454)
Cyclohexanone	108941		1*	4	U057	D	5000 (2270)
2-Cyclohexyl-4,6-dinitrophenol	131895	Phenol, 2-cyclohexyl-4,6-dinitro-	1*	4	P034	B	100 (45.4)
1,3-Cyclopentadiene, 1,2,3,4,5,5-hexachloro-	77474	Hexachlorocyclopentadiene	1	1,2,4	U130	A	10 (4.54)
Cyclophosphamide	50180	2H-1,3,2-Oxazaphosphorin-2-amine, N,N-bis(2-chloroethyl)tetrahydro-,2-oxide	1*	4	U058	A	10 (4.54)
2,4-D Acid	94757	Acetic acid (2,4-dichlorophenoxy)-2,4-D, salts and esters.	100	1,4	U240	B	100 (45.4)
2,4-D Ester	94111; 94791; 94804; 1320189; 1928387; 1928616; 1929733; 2971382; 25168267; 53467111		100	1		B	100 (45.4)
2,4-D, salts and esters	94757	Acetic acid (2,4-dichlorophenoxy)-2,4-D Acid	100	1,4	U240	B	100 (45.4)
Daunomycin	20830813	5,12-Naphthacenedione, 8-acetyl-10-[3-amino-2,3,6- trideoxy-alpha-L-lyxo-hexo-pyranosyl)oxy]-7,8,9,10- tetrahydro-6,8,11-trihydroxy-1-methoxy-, (8S-cis)-.	1*	4	U059	A	10 (4.54)
DDD	72548	Benzene, 1,1'-(2,2-dichloroethylidene)bis[4-chloro- ; TDE; 4,4' DDD	1	1,2,4	U060	X	1 (0.454)
4,4' DDD	72548	Benzene, 1,1'-(2,2-dichloroethylidene)bis[4-chloro- ; DDD; TDE	1	1,2,4	U060	X	1 (0.454)
DDE	72559	DDE	1*	2		X	1 (0.454)
4,4' DDE	72559	Benzene, 1,1'-(2,2-dichloroethylidene)bis[4-chloro- ; DDE	1*	2		X	1 (0.454)
DDT	50293	Benzene, 1,1'-(2,2,2-trichloroethylidene)bis[4-chloro- ; 4,4'DDT	1	1,2,4	U061	X	1 (0.454)
4,4'DDT	50293	Benzene, 1,1'-(2,2,2-trichloroethylidene)bis[4-chloro- ; DDT	1	1,2,4	U061	X	1 (0.454)

Hazardous substance	CASRN	Regulatory synonyms	Statutory RQ	Statutory Code†	RCRA waste Number	Final RQ Category	Final RQ Pounds (Kg)
DDT AND METABOLITES	N.A.		1*	2		B	100 (45.4)**
Diallate	2303164	Carbamothioic acid, bis(1-methylethyl)-, S-(2,3-dichloro-2-propenyl) ester.	1*	4	U062	X	1 (0.454)
Diazinon	333415		1	1		X	1 (0.454)
Dibenz[a,h]anthracene	53703	Dibenzo[a,h]anthracene; 1,2:5,6-Dibenzanthracene	1*	2,4	U063	X	1 (0.454)
1,2:5,6-Dibenzanthracene	53703	Dibenzo[a,h]anthracene	1*	2,4	U063	X	1 (0.454)
Dibenzo[a,h]anthracene	53703	1,2:5,6-Dibenzanthracene	1*	2,4	U063		
Dibenzo[a,j]pyrene	189559	Benzo[rst]pentaphene	1*	4	U064	A	10 (4.54)
1,2-Dibromo-3-chloropropane	96128	Propane, 1,2-dibromo-3-chloro-	1*	4	U066	X	1 (0.454)
Dibutyl phthalate	84742	n-Butyl phthalate; 1,2-Benzenedicarboxylic acid, dibutyl ester	100	1,2,4	U069	A	10 (4.54)
Di-n-butyl phthalate	84742	Dibutyl phthalate; n-Butyl phthalate; 1,2-Benzenedicarboxylic acid, dibutyl ester	100	1,2,4	U069	A	10 (4.54)
Dicamba	1918009		1000	1		C	1000 (454)
Dichlobenil	1194656		1000	1		B	100 (45.4)
Dichlone	117806		1	1		X	1 (0.454)
Dichlorobenzene	25321226		100	1,2,4			
1,2-Dichlorobenzene	95501	Benzene, 1,2-dichloro-; o-Dichlorobenzene	100	2,4	U070	B	100 (45.4)
1,3-Dichlorobenzene	541731	Benzene, 1,3-dichloro-; m-Dichlorobenzene	1*	1,2,4	U071	B	100 (45.4)
1,4-Dichlorobenzene	106467	Benzene, 1,4-dichloro-; p-Dichlorobenzene	100	2,4	U072	B	100 (45.4)
m-Dichlorobenzene	541731	Benzene, 1,3-dichloro 1,3-Dichlorobenzene	1*	1,2,4	U071	B	100 (45.4)
o-Dichlorobenzene	95501	Benzene, 1,2-dichloro 1,2-Dichlorobenzene	100	1,2,4	U070	B	100 (45.4)
p-Dichlorobenzene	106467	Benzene, 1,4-Dichlorobenzene	100	2	U072	B	100 (45.4)**
DICHLOROBENZIDINE	N.A.	[1,1'-Biphenyl]-4,4'diamine,3,3'dichloro-	1*	2,4			
3,3'-Dichlorobenzidine	91941		1*	2	U073	X	1 (0.454)
Dichlorobromomethane	75274		1*	4	U074	D	5000 (2270)
1,4-Dichloro-2-butene	764410	2-Butene, 1,4-dichloro-	1*	4	U075	X	1 (0.454)
Dichlorodifluoromethane	75718	Methane, dichlorodifluoro-	1*	2,4	U076	D	5000 (2270)
1,1-Dichloroethane	75343	Ethane, 1,1-dichloro-; Ethylidene dichloride	1*	1,2,4	U077	C	1000 (454)
1,2-Dichloroethane	107062	Ethane, 1,2-dichloro-; Ethylene dichloride	5000	1,2,4	U078	B	100 (45.4)
1,1-Dichloroethylene	75354	Ethene, 1,1-dichloro-; Vinylidene chloride	5000	1,2,4	U079	B	100 (45.4)
1,2-Dichloroethylene	156605	Ethene, 1,2-dichloro- (E)	1*	2,4		C	1000 (454)

Substance	CAS No.	Synonym(s)	Statutory RQ	RCRA Code No.	Waste Code	Category	Final RQ lb (kg)
Dichloroethyl ether	111444	Bis (2-chloroethyl) ether; Ethane, 1,1'-oxybis[2-chloro-	1*	2,4	U025	A	10 (4.54)
Dichloroisopropyl ether	108601	Propane, 2,2'-oxybis[2-chloro-	1*	2,4	U027	C	1000 (454)
Dichloromethoxy ethane	111911	Bis(2-chloroethoxy) methane; Ethane, 1,1'-[methylenebis(oxy)]bis[2-chloro-	1*	2,4	U024	C	1000 (454)
Dichloromethyl ether	542881	Methane, oxybis(chloro-	1*	4	P016	A	10 (4.54)
2,4-Dichlorophenol	120832	Phenol, 2,4-dichloro-	1*	2,4	U081	B	100 (45.4)
2,6-Dichlorophenol	87650	Phenol, 2,6-dichloro-	1*	4	U082	B	100 (45.4)
Dichlorophenylarsine	696286	Arsonous dichloride, phenyl-	1*	4	P036	X	1 (0.454)
Dichloropropane	26638197		5000	1		C	1000 (454)
1,1-Dichloropropane	78999						----
1,3-Dichloropropane	142289						----
1,2-Dichloropropane	78875	Propane, 1,2-dichloro-; Propylene dichloride	5000	1,2,4	U083	C	1000 (454)
Dichloropropane—Dichloropropene (mixture)	8003198		5000	1		B	100 (45.4)
Dichloropropene	26952238		5000	1		B	100 (45.4)
2,3-Dichloropropene	78886						
1,3-Dichloropropene	542756	1-Propene, 1,3-dichloro-	5000	1,2,4	U084	B	100 (45.4)
2,2-Dichloropropionic acid	75990		5000	1		D	5000 (2270)
Dichlorvos	62737		10	1		A	10 (4.54)
Dicofol	115322		5000	1		A	10 (4.54)
Dieldrin	60571	2,7:3,6-Dimethanonaphth[2,3-b]oxirene, 3,4,5,6,9,9-hexachloro-1a,2,2a,3,6,6a,7,7a-octahydro-, (1aalpha,2beta,2aalpha,3beta,6beta,6aalpha,7beta,7aalpha)-	1	1,2,4	P037	X	1 (0.454)
1,2:3,4-Diepoxybutane	1464535	2,2-Bioxirane	1*	4	U085	A	10 (4.54)
Diethylamine	109897		1000	1		B	100 (454.4)
Diethylarsine	692422	Arsine, diethyl-	1*	4	P038	X	1 (0.454)
1,4-Diethylenedioxide	123911	1,4-Dioxane	1*	4	U108	B	100 (45.4)
Diethylhexyl phthalate	117817	Bis (2-ethylhexyl)phthalate; 1,2-Benzenedicarboxylic acid, [bis(2-ethylhexyl)] ester	1*	2,4	U028	B	100 (45.4)
N,N-Diethylhydrazine	1615801	Hydrazine, 1,2-diethyl-	1*	4	U086	A	10 (4.54)
O,O-Diethyl S-methyl dithiophosphate	3288582	Phosphorodithioic acid, O,O-diethyl S-methyl ester	1*	4	U087	D	5000 (2270)
Diethyl-p-nitrophenyl phosphate	311455	Phosphoric acid, diethyl 4-nitrophenyl ester	1*	4	P041	B	100 (45.4)
Diethyl phthalate	84662	1,2-Benzenedicarboxylic acid, diethyl ester	1*	2,4	U088	C	1000 (454)
O,O-Diethyl O-pyrazinyl phosphorothioate	297972	Phosphorothioic acid, O,O-diethyl O-pyrazinyl ester	1*	4	P040	B	100 (45.4)
Diethylstilbestrol	56531	Phenol, 4,4'-(1,2-diethyl-1,2-ethenediyl)bis-, (E)	1*	4	U089	X	1 (0.454)
Dihydrosafrole	94586	1,3-Benzodioxole, 5-propyl-	1*	4	U090	A	10 (4.54)
Diisopropylfluorophosphate	55914	Phosphorofluoridic acid, bis(1-methylethyl) ester	1*	4	P043	B	100 (45.4)
1,4,5,8-Dimethanonaphthalene, 1,2,3,4,10,10-hexachloro-1,4,4a,5,8,8a-hexahydro-, (1alpha,4alpha,4abeta,5alpha,8alpha,8abeta)-	309002	Aldrin	1	1,2,4	P004	X	1 (0.454)
Isodrin	465736		1*	4	P060	X	*1 (0.454)
Dieldrin	60571	2,7:3,6-Dimethanonaphth[2,3-b]oxirene, 3,4,5,6,9,9-hexachloro-1a,2,2a,3,6,6a,7,7a-octahydro-, (1aalpha,2beta,2aalpha,3beta,6beta,6aalpha)-	1	1,2,4	P037	X	1 (0.454)

Hazardous substance	CASRN	Regulatory synonyms	RQ (Statutory)	Code† (Statutory)	RCRA waste Number (Statutory)	Category (Final RQ)	Pounds (Kg) (Final RQ)
6aalpha,7beta,7aalpha)-2,7,3,6- 3,4,5,6,9,9-Dimethanonaphth[2,3-b]oxirene, hexachloro-1a,2,2a,3,6,6a,7,7a-octa-hydro, (1aalpha,2beta,2abeta,3alpha,3alpha,6alpha, 6abeta,7beta,7aalpha)-	72208	Endrin / Endrin, & metabolites	1	1,2,4	P051	X	1 (0.454)
Dimethoate	60515	Phosphorodithioic acid, O,O-dimethyl S-[2(methylamino)-2-oxoethyl] ester.	1*	4	P044	A	10 (4.54)
3,3'-Dimethoxybenzidine	119904	[1,1'-Biphenyl]-4,4'diamine,3,3'dimethoxy-	1*	4	U091	B	100 (45.4)
Dimethylamine	124403	Methanamine, N-methyl-	1000	1,4	U092	C	1000 (454)
p-Dimethylaminoazobenzene	60117	Benzenamine, N,N-dimethyl-4-(phenylazo)-	1*	4	U093	A	10 (4.54)
7,12-Dimethylbenz[a]anthracene	57976	Benz[a]anthracene, 7,12-dimethyl-	1*	4	U094	X	1 (0.454)
3,3'-Dimethylbenzidine	119937	[1,1'Biphenyl]-4,4'-diamine,3,3'-dimethyl-	1*	4	U095	A	10 (4.54)
alpha,alpha-Dimethylbenzylhydroperoxide	80159	Hydroperoxide, 1-methyl-1-phenylethyl-	1*	4	U096	A	10 (4.54)
Dimethylcarbamoyl chloride	79447	Carbamic chloride, dimethyl-	1*	4	U097	X	1 (0.454)
1,1-Dimethylhydrazine	57147	Hydrazine, 1,1-dimethyl-	1*	4	U098	A	10 (4.54)
1,2-Dimethylhydrazine	540738	Hydrazine, 1,2-dimethyl-	1*	4	U099	X	1 (0.454)
alpha,alpha-Dimethylphenethylamine	122098	Benzeneethanamine, alpha,alpha-dimethyl-	1*	4	P046	D	5000 (2270)
2,4-Dimethylphenol	105679	Phenol, 2,4-dimethyl-	1*	2,4	U101	D	5000 (2270)
Dimethyl phthalate	131113	1,2-Benzenedicarboxylic acid, dimethyl ester	1*	2,4	U102	B	100 (45.4)
Dimethyl sulfate	77781	Sulfuric acid, dimethyl ester	1000	1	U103	B	100 (45.4)
Dinitrobenzene (mixed)	25154545						
m-Dinitrobenzene	99650						
o-Dinitrobenzene	528290						
p-Dinitrobenzene	100254						
4,6-Dinitro-o-cresol and salts	534521	Phenol, 2-methyl-4,6-dinitro-	1*	2,4	P047	A	10 (4.54)
Dinitrophenol	25550587		1000	1		A	10 (4.54)
2,5-Dinitrophenol	329715						
2,6-Dinitrophenol	573568						
2,4-Dinitrophenol	51285	Phenol, 2,4-dinitro-	1000	1,2,4	P048	A	10 (4.54)
Dinitrotoluene	25321146		1000	1,2		A	10 (4.54)
3,4-Dinitrotoluene	610399						
2,4-Dinitrotoluene	121142	Benzene, 1-methyl-2,4-dinitro-	1000	1,2,4	U105	A	10 (4.54)
2,6-Dinitrotoluene	606202	Benzene, 2-methyl-1,3-dinitro-	1000	1,2,4	U106	B	100 (45.4)
Dinoseb	88857	Phenol, 2-(1-methylpropyl)-4,6-dinitro	1*	4	P020	C	1000 (454)
Di-n-octyl phthalate	117840	1,2-Benzenedicarboxylic acid, dioctyl ester	1*	2,4	U107	D	5000 (2270)
1,4-Dioxane	123911	1,4-Diethylenedioxide	1*	4	U108	B	100 (45.4)
DIPHENYLHYDRAZINE	N.A.		1*	2		A	10 (4.54)
1,2-Diphenylhydrazine	122667	Hydrazine, 1,2-diphenyl-	1*	2,4	U109	A	10 (4.54)
Diphosphoramide, octamethyl-	152169	Octamethylpyrophosphoramide	1*	4	P085	B	100 (45.4)
Diphosphoric acid, tetraethyl ester	107493	Tetraethyl pyrophosphate	100	1,4	P111	A	10 (4.54)
Dipropylamine	142847	1-Propanamine, N-propyl-	1*	1,4	U110	D	5000 (2270)
Di-n-propylnitrosamine	621647	1-Propanamine, N-nitroso-N-propyl-	1*	2,4	U111	A	10 (4.54)

Hazardous Substance	CASRN	Regulatory Synonyms	Statutory RQ	Code	RCRA Waste No.	Final RQ Category	Final RQ Pounds (Kg)
Diquat	85007, 2764729		1000	1		C	1000 (454)
Disulfoton	298044	Phosphorodithioic acid, o,o-diethyl S-[2-(ethylthio)ethyl]ester.	1	1,4	P039	X	1 (0.454)
Dithiobiuret	541537	Thioimidodicarbonic diamide ([H2N]C(S))2NH	1*	4	P049	B	100 (45.4)
Diuron	330541		100	1		B	100 (45.4)
Dodecylbenzenesulfonic acid	27176870		1000			C	1000 (454)
Endosulfan	115297	6,9-Methano-2,4,3-benzodioxathiepin, 6,7,8,9,10,10-hexachloro-1,5,5a,6,9,9a-hexahydro-, 3-oxide.	1	1,2,4	P050	X	1 (0.454)
alpha - Endosulfan	959988		1*	2		X	1 (0.454)
beta - Endosulfan	33213669		1*	2		X	1 (0.454)**
ENDOSALFAN AND METABOLITES	N.A.		1*				
Endosulfan sulfate	1031078		1	2		X	1 (0.454)
Endothall	145733	7-Oxabicyclo[2.2.1]heptane-2,3-dicarboxylic acid	1*	4	P088	C	1000 (454)
Endrin	72208	Endrin, & metabolites 2,7:3,6-Dimethanonaphth[2,3-b]oxirene, 3,4,5,6,9,9 -hexachloro-1a,2,2a,3, 6,6a,7,7a-octa-hydro-, (1aalpha, 2beta,2abeta,3alpha,6alpha, 6abeta,7beta,7aalpha)-	1	1,2,4	P051	X	1 (0.454)
Endrin aldehyde	7421934		1*	2		X	1 (0.454)**
ENDRIN AND METABOLITES	N.A.		1*				
Endrin, & metabolites	72208	Endrin 2,7:3,6-Dimethanonaphth[2,3-b]oxirene, 3,4,5,6,9,9-hexachloro-1a,2,2a,3, 6,6a,7,7a-octa-hydro-, (1aalpha, 2beta,2abeta,3alpha,6alpha, 6abeta,7beta,7aalpha)-	1	1,2,4	P051	X	1 (0.454)
Epichlorohydrin	106898	Oxirane, (chloromethyl)-	1000	1,4	U041	B	100 (45.4)
Epinephrine	51434	1,2-Benzenediol,4-[1-hydroxy-2-(methylamino)ethyl]-.	1*	4	P042	C	1000 (454)
Ethanal	75070	Acetaldehyde	1000	1,4	U001	C	1000 (454)
Ethanamine, N-ethyl-N-nitroso-	55185	N-Nitrosodiethylamine	1*	4	U174	X	1 (0.454)
1,2-Ethanediamine, N,N-dimethyl-N'-2-pyridinyl-N'-(2-thienylmethyl)-	91805	Methapyrilene	1*	4	U155	D	5000 (2270)
Ethane, 1,2-dibromo-	106934	Ethylene dibromide	1000	1,4	U067	X	1 (0.454)
Ethane, 1,1-dichloro-	75343	Ethylidene dichloride; 1,1-Dichloroethane	1*	2,4	U076	C	1000 (454)
Ethane, 1,2-dichloro-	107062	Ethylene dichloride; 1,2-Dichloroethane	5000	1,2,4	U077	B	100 (45.4)
Ethanedinitrile	460195	Cyanogen	1*	4	P031	B	100 (45.4)
Ethane, hexachloro-	67721	Hexachloroethane	1*	2,4	U131	B	100 (45.4)
Ethane, 1,1'-[methylenebis(oxy)]bis(2-chloro-	111911	Bis(2-chloroethoxy) methane; Dichloromethoxy ethane	1*	2,4	U024	C	1000 (454)
Ethane, 1,1'-oxybis-	60297	Ethyl ether	1*	4	U117	B	100 (45.4)
Ethane, 1,1'-oxybis[2-chloro-	111444	Bis (2-chloroethyl) ether; Dichloroethyl ether	1*	2,4	U025	A	10 (4.54)
Ethane, pentachloro-	76017	Pentachloroethane	1*	4	U184	A	10 (4.54)

Hazardous substance	CASRN	Regulatory synonyms	Statutory RQ	Statutory Code†	RCRA waste Number	Final RQ Category	Final RQ Pounds (Kg)
Ethane, 1,1,1,2-tetrachloro-	630206	1,1,1,2-Tetrachloroethane	1*	4	U208	B	100 (45.4)
Ethane, 1,1,2,2-tetrachloro-	79345	1,1,2,2-Tetrachloroethane	1*	2,4	U209	B	100 (45.4)
Ethanethioamide	62555	Thioacetamide	1*	4	U218	A	10 (4.54)
Ethane, 1,1,1-trichloro-	71556	Methyl chloroform	1*	2,4	U226	C	1000 (454)
Ethane, 1,1,2-trichloro-	79005	1,1,2-Trichloroethane	1*	2,4	U227	B	100 (45.4)
Ethanimidothioic acid, N-[[(methyl-amino)carbonyl]oxy]-, methyl ester	16752775	Methomyl	1*	4	P066	B	100 (45.4)
Ethanol, 2-ethoxy-	110805	Ethylene glycol monoethyl ether	1*	4	U359	C	1000 (454)
Ethanol, 2,2'-(nitrosoimino)bis-	1116547	N-Nitrosodiethanolamine	1*	4	U173	X	1 (0.454)
Ethanone, 1-phenyl-	98862	Acetophenone	1*	2,3,4	U004	D	5000 (2270)
Ethene, chloro-	75014	Vinyl chloride	1*	2,4	U043	X	1 (0.454)
Ethene, 2-chloroethoxy-	110758	2-Chloroethyl vinyl ether	1*	2,4	U042	C	1000 (454)
Ethene, 1,1-dichloro-	75354	Vinylidene chloride	5000	1,2,4	U078	B	100 (45.4)
Ethene, 1,2-dichloro- (E)	156605	1,1-Dichloroethylene	1*	2,4	U079	C	1000 (454)
Ethene, tetrachloro-	127184	Perchloroethylene; Tetrachloroethene; Tetrachloroethylene	1*	2,4	U210	B	100 (45.4)
Ethene, trichloro-	79016	Trichloroethene; Trichloroethylene	1000	1,2,4	U228	B	100 (45.4)
Ethion	563122		10	1		A	10 (4.54)
Ethyl acetate	141786	Acetic acid, ethyl ester	1*	4	U112	D	5000 (2270)
Ethyl acrylate	140885	2-Propenoic acid, ethyl ester	1*	4	U113	C	1000 (454)
Ethylbenzene	100414		1000	1,2		C	1000 (454)
Ethyl carbamate (urethane)	51796	Carbamic acid, ethyl ester	1*	4	U238	B	100 (45.4)
Ethyl cyanide	107120	Propanenitrile	1*	4	P101	A	10 (4.54)
Ethylenebisdithiocarbamic acid, salts & esters	111546	Carbamodithioic acid, 1,2-ethanediylbis, salts & esters.	1*	4	U114	D	5000 (2270)
Ethylenediamine	107153		1000	1		D	5000 (2270)
Ethylenediamine-tetraacetic acid (EDTA)	60004		5000	1		D	5000 (2270)
Ethylene dibromide	106934	Ethane, 1,2-dibromo-	1000	1,4	U067	X	1 (0.454)
Ethylene dichloride	107062	Ethane, 1,2-dichloro-; 1,2-Dichloroethane	5000	1,2,4	U077	B	100 (45.4)
Ethylene glycol monoethyl ether	110805	Ethanol, 2-ethoxy-	1*	4	U359	C	1000 (454)
Ethylene oxide	75218	Oxirane	1*	4	U115	A	10 (4.54)
Ethylenethiourea	96457	2-Imidazolidinethione	1*	4	U116	A	10 (4.54)
Ethyleneimine	151564	Aziridine	1*	4	P054	X	1 (0.454)
Ethyl ether	60297	Ethane, 1,1'-oxybis-	1*	4	U117	B	100 (45.4)
Ethylidene dichloride	75343	Ethane, 1,1-dichloro-; 1,1-Dichloroethane	1*	2,4	U076	C	1000 (454)

Hazardous Substance	CAS No.	Statutory RQ	Code	RCRA Waste No.	Final RQ Category	Final RQ lbs (Kg)
Ethyl methacrylate	97632	1*	4	U118	C	1000 (454)
Ethyl methanesulfonate	62500	1*	4	U119	X	1 (0.454)
Famphur	52857	1*	4	P097	C	1000 (454)
Phosphorothioic acid, O,[4-(di- methylamino) sulfonyl] phenyl] O,O-dimethyl ester.						
Ferric ammonium citrate	1185575	1000	1		C	1000 (454)
Ferric ammonium oxalate	2944674	1000	1		C	1000 (454)
	55488874					
Ferric chloride	7705080	1000	1		C	1000 (454)
Ferric fluoride	7783508	100	1		B	100 (45.4)
Ferric nitrate	10421484	1000	1		C	1000 (454)
Ferric sulfate	10028225	1000	1		C	1000 (454)
Ferrous ammonium sulfate	10045893	1000	1		C	1000 (454)
Ferrous chloride	7758943	100	1		B	100 (45.4)
Ferrous sulfate	7720787	1000	1		C	1000 (454)
	7782630					
Fluoranthene (Benzo[j,k]fluorene)	206440	1*	2,4	U120	B	100 (45.4)
Fluorene	86737	1*	2		D	5000 (2270)
Fluorine	7782414	1*	4	P056	A	10 (4.54)
Fluoroacetamide (Acetamide, 2-fluoro-)	640197	1*	4	P057	B	100 (45.4)
Fluoroacetic acid, sodium salt (Acetic acid, fluoro-, sodium salt)	62748	1*	4	P058	A	10 (4.54)
Formaldehyde	50000	1000	1,4	U122	B	100 (45.4)
Formic acid	64186	5000	1,4	U123	D	5000 (2270)
Mercury fulminate (Fulminic acid, mercury(2+)salt)	628864	1*	4	P065	A	10 (4.54)
Fumaric acid	110178	5000	1		D	5000 (2270)
Furan	110009	1*	4	U124	B	100 (45.4)
Furan, tetrahydro- (Tetrahydrofuran)	109999	1*	4	U213	C	1000 (454)
2-Furancarboxaldehyde (Furfural)	98011	1*	1,4	U125	D	5000 (2270)
2,5-Furandione (Maleic anhydride)	108316	1000	1,4	U147	D	5000 (2270)
Furfural (2-Furancarboxaldehyde)	98011	5000	1,4	U125	D	5000 (2270)
Furfuran (Furan)	110009	1000	4	U124	B	100 (45.4)
Streptozotocin (D-Glucose, 2-deoxy-2-[[(methylnitrosoamino)-carbonyl]amino])	18883664	1*	4	U206	X	1 (0.454)
Glucopyranose, 2-deoxy-2-[(3-methyl-3-nitrosoureido)-; D-Glucose, 2-deoxy-2-[[(methylnitrosoamino)-carbonyl]amino]-	18883664	1*	4	U206	X	1 (0.454)
Glycidylaldehyde (Oxiranecarboxyaldehyde)	765344	1*	4	U126	A	10 (4.54)
Guanidine, N-methyl-N'-nitro-N-nitroso- (MNNG)	70257	1*	4	U163	A	10 (4.54)
Guthion	86500	1*	1		X	1 (0.454)
HALOETHERS	N.A.	1*	2			**
HALOMETHANES	N.A.	1*	2			**
Heptachlor (4,7-Methano-1H-indene, 1,4,5,6,7,8,8-heptachloro-3a,4,7,7a-tetrahydro-)	76448	1	1,2,4	P059	X	1 (0.454)
HEPTACHLOR AND METABOLITES	N.A.		2			**
Heptachlor epoxide	1024573	1*	2		X	1 (0.454)
Hexachlorobenzene (Benzene, hexachloro-)	118741	1*	2,4	U127	A	10 (4.54)
Hexachlorobutadiene (1,3-Butadiene, 1,1,2,3,4,4-hexachloro-)	87683	1*	2,4	U128	X	1 (0.454)**
HEXACHLOROCYCLOHEXANE (all isomers)	608731	1*	2			**
Hexachlorocyclohexane (gamma isomer) (Cyclohexane, 1,2,3,4,5,6-hexachloro-, (1alpha,2alpha,3beta,4alpha,5alpha,6beta)-; gamma-BHC; Lindane)	58899	1	1,2,4	U129	X	1 (0.454)

Hazardous substance	CASRN	Regulatory synonyms	Statutory RQ	Statutory Code†	RCRA waste Number	Final RQ Category	Final RQ Pounds (Kg)
Hexachlorocyclopentadiene	77474	1,3-Cyclopentadiene,1,2,3,4,5,5-hexachloro-	1	1,2,4	U130	A	10 (4.54)
Hexachloroethane	67721	Ethane, hexachloro-	1*	2,4	U131	B	100 (45.4)
Hexachlorophene	70304	Phenol, 2,2'-methylenebis[3,4,6-trichloro-	1*	4	U132	B	100 (45.4)
Hexachloropropene	1888717	1-Propene, 1,1,2,3,3,3-hexachloro-	1*	4	U243	C	1000 (454)
Hexaethyl tetraphosphate	757584	Tetraphosphoric acid, hexaethyl ester	1*	4	P062	X	100 (45.4)
Hydrazine	302012		1*	4	U133	X	1 (0.454)
Hydrazine, 1,2-diethyl-	1615801	N,N'-Diethylhydrazine	1*	4	U086	A	10 (4.54)
Hydrazine, 1,1-dimethyl-	57147	1,1-Dimethylhydrazine	1*	4	U098	A	10 (4.54)
Hydrazine, 1,2-dimethyl-	540738	1,2-Dimethylhydrazine	1*	4	U099	X	1 (0.454)
Hydrazine, 1,2-diphenyl-	122667	1,2-Diphenylhydrazine	1*	2,4	U109	A	10 (4.54)
Hydrazine, methyl-	60344	Methyl hydrazine	1*	4	P068	A	10 (4.54)
Hydrazinecarbothioamide	79196	Thiosemicarbazide	1*	4	P116	B	100 (45.4)
Hydrochloric acid	7647010	Hydrogen chloride	5000	1		D	5000 (2270)
Hydrocyanic acid	74908	Hydrogen cyanide	10	1,4	P063	A	10 (4.54)
Hydrofluoric acid	7664393	Hydrogen fluoride	5000	1,4	U134	B	100 (45.4)
Hydrogen chloride	7647010	Hydrochloric acid	5000	1		D	5000 (2270)
Hydrogen cyanide	74908	Hydrocyanic acid	10	1,4	P063	A	10 (4.54)
Hydrogen fluoride	7664393	Hydrofluoric acid	5000	1,4	U134	B	100 (45.4)
Hydrogen sulfide H2S	7783064	Hydrogen sulfide H2S	100	1,4	U135	B	100 (45.4)
Hydrogen sulfide	7783064	Hydrogen sulfide	100	1,4	U135	B	100 (45.4)
Hydroperoxide, 1-methyl-1-phenylethyl-	80159	alpha,alpha-Dimethylbenzylhydroperoxide	1*	4	U096	A	10 (4.54)
2-Imidazolidinethione	96457	Ethylenethiourea	1*	4	U116	A	10 (4.54)
Indeno(1,2,3-cd)pyrene	193395	1,10-(1,2-Phenylene)pyrene	1*	2,4	U137	B	100 (45.4)
1,3-Isobenzofurandione	85449	Phthalic anhydride	1*	4	U190	D	5000 (2270)
Isobutyl alcohol	78831	1-Propanol, 2-methyl-	1*	4	U140	D	5000 (2270)
Isodrin	465736	1,4,5,8-Dimethanonaphthalene, 1,2,3,4,10,10-hexachloro-1,4,4a,5,8,8a-hexahydro-, (1alpha,4alpha,4abeta,5beta,8beta,8abeta)-	1*	4	P060	X	1 (0.454)
Isophorone	78591		1*	2		D	5000 (2270)
Isoprene	78795		1000	1		B	100 (45.4)
Isopropanolamine dodecylbenzenesulfonate	42504461		1000	1		C	1000 (454)
Isosafrole	120581	1,3-Benzodioxole, 5-(1-propenyl)-	1*	4	U141	B	100 (45.4)
3(2H)-Isoxazolone, 5-(aminomethyl)-	2763964	Muscimol ...... 5-(Aminomethyl)-3-isoxazolol	1*	4	P007	C	1000 (454)
Kepone	143500	1,3,4-Metheno-2H-cyclobuta[cd]pentalen-2-one, 1,1a,3,3a,4,5,5,5a,5b,6-decachlorooctahydro-	1	1,4	U142	X	1 (0.454)
Lasiocarpine	303344	2-Butenoic acid, 2-methyl-, 7[[2,3-dihydroxy-2-(1-methoxyethyl)-3-methyl-1-oxobutoxy]methyl]-2,3,5,7a-tetrahydro-1H-pyrrolizin-1-yl ester, [1S-[1alpha(Z), 7(2S*,3R*),7aalpha]]-	1*	4	U143	A	10 (4.54)
Lead††	7439921		1*	2		A	10 (4.54)

Substance	CAS No.	Synonym / Chemical name	Code	Waste No.	Cat.	Statutory RQ	Final RQ pounds (kilograms)
Lead acetate	301042	Acetic acid, lead(2+) salt	1,4	U144	A	5000	10 (4.54)
LEAD AND COMPOUNDS	N.A.		2		X	1*	1 (0.454)**
Lead arsenate	7784409 7645252 10102484		1			5000	
Lead, bis(acetato-O)tetrahydroxytri-	1335326	Lead subacetate	4	U146	A	1*	10 (4.54)
Lead chloride	7758954		1		A	5000	10 (4.54)
Lead fluoborate	13814965		1		A	5000	10 (4.54)
Lead fluoride	7783462		1		A	1000	10 (4.54)
Lead iodide	10101630		1		A	5000	10 (4.54)
Lead nitrate	10099748		1		A	5000	10 (4.54)
Lead phosphate	7446277	Phosphoric acid, lead(2+) salt (2:3)	4	U145	A	1*	10 (4.54)
Lead stearate	1072351 7428480 52652592 56189094		1		A	5000	10 (4.54)
Lead subacetate	1335326	Lead, bis(acetato-O)tetrahydroxytri-	4	U146	A	1*	10 (4.54)
Lead sulfate	7446142 15739807		1		A	5000	10 (4.54)
Lead sulfide	1314870		1		A	5000	10 (4.54)
Lead thiocyanate	592870		1		A	5000	10 (4.54)
Lindane	58899	Cyclohexane, 1,2,3,4,5,6-hexachloro-, (1alpha,2alpha,3beta,4alpha,5alpha,6beta)- gamma-BHC. Hexachlorocyclohexane (gamma isomer)	1,2,4	U129	X	1	1 (0.454)
Lithium chromate	14307358		1		A	1000	10 (4.54)
Malathion	121755		1		B	10	100 (45.4)
Maleic acid	110167		1		D	5000	5000 (2270)
Maleic anhydride	108316	2,5-Furandione	1,4	U147	D	5000	5000 (2270)
Maleic hydrazide	123331	3,6-Pyridazinedione, 1,2-dihydro-	4	U148	D	5000	1000 (454)
Malononitrile	109773	Propanedinitrile	4	U149	C	1*	1 (0.454)
Melphalan	148823	L-Phenylalanine, 4-[bis(2-chloroethyl) amino]	4	U150	X	1*	10 (4.54)
Mercaptodimethur	2032657		1		X	100	1 (0.454)
Mercuric cyanide	592041		1		A	1	10 (4.54)
Mercuric nitrate	10045940		1		A	10	10 (4.54)
Mercuric sulfate	7783359		1		A	10	10 (4.54)
Mercuric thiocyanate	592858		1		A	10	10 (4.54)
Mercurous nitrate	10415755 7782867		1		A	10	10 (4.54)
MERCURY AND COMPOUNDS	7439976 N.A.		2,3,4	U151	X	1*	1 (0.454)
Mercury, (acetate-O)phenyl-	62384	Phenylmercury acetate	2	P092	B	1*	100 (45.4)
Mercury fulminate	628864	Fulminic acid, mercury(2+)salt	1	P065	B	1*	10 (4.54)
Methacrylonitrile	126987	2-Propenenitrile, 2-methyl-	4	U152	C	1*	1000 (454)
Methanamine, N-methyl-	124403	Dimethylamine	1,4	U092	C	1000	1000 (454)
Methanamine, N-methyl-N-nitroso-	62759	N-Nitrosodimethylamine	2,4	P082	C	1*	10 (4.54)
Methane, bromo-	74839	Methyl bromide	2,4	U029	C	1*	1000 (454)
Methane, chloro-	74873	Methyl chloride	2,4	U045	C	1*	100 (45.4)
Methane, chloromethoxy-	107302	Chloromethyl methyl ether	4	U046	B	1*	10 (4.54)
Methane, dibromo-	74953	Methylene bromide	4	U068	C	1*	1000 (454)

Hazardous substance	Regulatory synonyms	CASRN	Statutory			Final RQ	
			RQ	Code †	RCRA waste Number	Category	Pounds (Kg)
Methane, dichloro-	Methylene chloride	75092	1*	2,4	U080	C	1000 (454)
Methane, dichlorodifluoro-	Dichlorodifluoromethane	75718	1*	4	U075	D	5000 (2270)
Methane, iodo-	Methyl iodide	74884	1*	4	U138	B	100 (45.4)
Methane, isocyanato-	Methyl isocyanate	624839	1*	3,4	P064	A	10 (4.54)
Methane, oxybis(chloro-	Dichloromethyl ether	542881	1*	4	P016	A	10 (4.54)
Methanesulfenyl chloride, trichloro-	Trichloromethanesulfenyl chloride	594423	1*	4	P118	B	100 (45.4)
Methanesulfonic acid, ethyl ester	Ethyl methanesulfonate	62500	1*	4	U119	X	1 (0.454)
Methane, tetrachloro-	Carbon tetrachloride	56235	5000	1,2,4	U211	A	10 (4.54)
Methane, tetranitro-	Tetranitromethane	509148	1*	4	P112	A	10 (4.54)
Methane, tribromo-	Bromoform	75252	1*	2,4	U225	B	100 (45.4)
Methane, trichloro-	Chloroform	67663	5000	1,2,4	U044	A	10 (4.54)
Methane, trichlorofluoro-	Trichloromonofluoromethane	75694	1*	4	U121	D	5000 (2270)
Methanethiol	Thiomethanol	74931	100	1,4	U153	B	100 (45.4)
6,9-Methano-2,4,3-benzodioxathiepin, 6,7,8,9,10,10-hexachloro-1,5,5a,6,9,9a- hexahydro-, 3-oxide	Endosulfan	115297	1	1,2,4	P050	X	1 (0.454)
1,3,4-Metheno-2H-cyclobuta[cd]pentalen-2-one, 1,1a,3,3a,4,5,5,5a,5b,6-decachlorooctahydro-	Kepone	143500	1	1,4	U142	X	1 (0.454)
4,7-Methano-1H-indene, 1,4,5,6,7,8,8-heptachloro-3a,4,7,7a-tetrahydro-	Heptachlor	76448	1	1,2,4	P059	X	1 (0.454)
4,7-Methano-1H-indene, 1,2,4,5,6,7,8,8-octachloro-2,3,3a,4,7,7a-hexahydro-	Chlordane, alpha & gamma isomers Chlordane, technical	57749	1	1,2,4	U036	X	1 (0.454)
Methanol	Methyl alcohol	67561	1*	4	U154	D	5000 (2270)
Methapyrilene	1,2-Ethanediamine, N,N-dimethyl-N'-2-pyridinyl-N'-(2-thienylmethyl)-	91805	1*	4	U155	D	5000 (2270)
Methomyl	Ethanimidothioic acid, N-[[(methyl-amino)carbonyl]oxy]-, methyl ester.	16752775	1*	4	P066	B	100 (45.4)
Methoxychlor	Benzene, 1,1'-(2,2,2-trichloroethylidene) bis[4-methoxy-.	72435	1	1,4	U247	X	1 (0.454)
Methyl alcohol	Methanol	67561	1*	4	U154	D	5000 (2270)
Methyl bromide	Methane, bromo-	74839	1*	2,4	U029	C	1000 (454)
1-Methylbutadiene	1,3-Pentadiene	504609	1*	4	U186	B	100 (45.4)
Methyl chloride	Methane, chloro-	74873	1*	2,4	U045	B	100 (45.4)
Methyl chlorocarbonate	Carbonochloridic acid, methyl ester	79221	1*	4	U156	C	1000 (454)
Methyl chloroform	Ethane, 1,1,1-trichloro- 1,1,1-Trichloroethane	71556	1*	2,4	U226	C	1000 (454)
Methyl chloroformate	Carbonochloridic acid, methyl ester	79221	1*	4	U156	C	1000 (454)
3-Methylcholanthrene	Benz[j]aceanthrylene, 1,2-dihydro-3-methyl-	56495	1*	4	U157	A	10 (4.54)

Substance	CAS No.	Description		Code		Waste Code	RQ
4,4'-Methylenebis(2-chloroaniline)	101144	Benzenamine, 4,4'-methylenebis(2-chloro-	1*	4	U158	A	10 (4.54)
Methylene bromide	74953	Methane, dibromo-	1*	4	U068	C	1000 (454)
Methylene chloride	75092	Methane, dichloro-	1*	2,4	U080	C	1000 (454)
Methyl ethyl ketone (MEK)	78933	2-Butanone	1*	4	U159	D	5000 (2270)
Methyl ethyl ketone peroxide	1338234	2-Butanone peroxide	1*	4	U160	A	10 (4.54)
Methyl hydrazine	60344	Hydrazine, methyl-	1*	4	P068	A	10 (4.54)
Methyl iodide	74884	Methane, iodo-	1*	4	U138	B	100 (45.4)
Methyl isobutyl ketone	108101	4-Methyl-2-pentanone	1*	4	U161	D	5000 (2270)
Methyl isocyanate	624839	Methane, isocyanato-	1*	3,4	P064	A	10 (4.54)
2-Methyllactonitrile	75865	Acetone cyanohydrin	10	1,4	P069	A	10 (4.54)
		Propanenitrile, 2-hydroxy-2-methyl-					
Methylmercaptan	74931	Methanethiol	100	1,4	U153	B	100 (45.4)
		Thiomethanol					
Methyl methacrylate	80626	2-Propenoic acid, 2-methyl-, methyl ester	5000	1,4	U162	C	1000 (454)
Methyl parathion	298000	Phosphorothioic acid, O,O-dimethyl O-(4-nitrophenyl) ester.	100	1,4	P071	B	100 (45.4)
4-Methyl-2-pentanone	108101	Methyl isobutyl ketone	1*	4	U161	D	5000 (2270)
Methylthiouracil	56042	4(1H)-Pyrimidinone, 2,3-dihydro-6-methyl-2-thioxo-	1*	1	U164	A	10 (4.54)
Mevinphos	7786347		1	1		A	10 (4.54)
Mexacarbate	315184		1000	1		C	1000 (454)
Mitomycin C	50077	Azirino[2',3':3,4]pyrrolo[1,2-a]indole-4,7-dione,6-amino-8-[[(aminocarbonyl)oxy] methyl]-1,1a,2,8,8a,8b-hexahydro-8a-methoxy-5-methyl-, [1aS-(1aalpha,8beta,8aalpha,8aalpha,8balpha)]-	1*	4	U010	A	10 (4.54)
MNNG	70257	Guanidine, N-methyl-N'-nitro-N-nitroso-	1*	4	U163	A	10 (4.54)
Monoethylamine	75047		1000	1		B	100 (45.4)
Monomethylamine	74895		1000	1		B	100 (45.4)
Multi Source Leachate			4	4	F039	X	1 (0.454)
Muscimol	2763964	3(2H)-Isoxazolone, 5-(aminomethyl)- (Aminomethyl)-3-isoxazolol.	1*	4	P007	C	1000 (454)
Naled	300765		10	1		A	10 (4.54)
5,12-Naphthacenedione, 8-acetyl-10-[3-amino-2,3,6-trideoxy-alpha-L-lyxo-hexopyranosyl)oxy]-7,8,9,10-tetrahydro-6,8,11-trihydroxy-1-methoxy-, (8S-cis)-	20830813	Daunomycin	1*	4	U059	A	10 (4.54)
1-Naphthalenamine	134327	alpha-Naphthylamine	1*	4	U167	B	100 (45.4)
2-Naphthalenamine	91598	beta-Naphthylamine	1*	4	U168	A	10 (4.54)
Naphthalenamine, N,N'-bis(2-chloroethyl)-	494031	Chlornaphazine	1*	4	U026	B	100 (45.4)
Naphthalene	91203		5000	1,2,4	U165	B	100 (45.4)
Naphthalene, 2-chloro-	91587	beta-Chloronaphthalene 2-Chloronaphthalene	1*	2,4	U047	D	5000 (2270)
1,4-Naphthalenedione	130154	1,4-Naphthoquinone	1*	4	U166	D	5000 (2270)
2,7-Naphthalenedisulfonic acid, 3,3'-[(3,3'-dimethyl-(1,1'-biphenyl)-4,4'-diyl)-bis(azo)]bis(5-amino-4-hydroxy)-tetrasodium salt.	72571	Trypan blue	1*	4	U236	A	10 (4.54)
Naphthenic acid	1338245		100	1	U166	B	100 (45.4)
1,4-Naphthoquinone	130154	1,4-Naphthoquinone	1*	4	U167	D	5000 (2270)
alpha-Naphthylamine	134327	1-Naphthalenamine	1*	4	U168	A	10 (4.54)
beta-Naphthylamine	91598	2-Naphthalenamine	1*	4	P072	B	100 (45.4)
alpha-Naphthylthiourea	86884	Thiourea, 1-naphthalenyl-	1*	4		B	100 (45.4)
Nickel††	7440020		1*	2		B	100 (45.4)
Nickel ammonium sulfate	15699180		5000	1		B	100 (45.4)

Hazardous substance	CASRN	Regulatory synonyms	Statutory RQ	Statutory Code †	RCRA waste Number	Final RQ Category	Final RQ Pounds (Kg)
NICKEL AND COMPOUNDS	N.A.						**
Nickel carbonyl	13463393	Nickel carbonyl Ni(CO)4, (T-4)-	1*	2	P073	A	10 (4.54)
Nickel carbonyl Ni(CO)4, (T-4)-	13463393	Nickel carbonyl	1*	4	P073	A	10 (4.54)
Nickel chloride	7718549		5000	1		B	100 (45.4)
	37211055						
Nickel cyanide	557197	Nickel cyanide Ni(CN)2	1*	4	P074	A	10 (4.54)
Nickel cyanide Ni(CN)2	557197	Nickel cyanide	1*	4	P074	A	10 (4.54)
Nickel hydroxide	12054487		1000	1		A	10 (4.54)
Nickel nitrate	14216752		5000	1		B	100 (45.4)
Nickel sulfate	7786814		5000	1		B	100 (45.4)
Nicotine, & salts	54115	Pyridine, 3-(1-methyl-2-pyrolidinyl)-, (S)-	1000	4	P075	B	100 (45.4)
Nitric acid	7697372		1000	1		C	1000 (454)
Nitric acid, thallium (1+) salt	10102451	Thallium (I) nitrate	1*	4	U217	B	100 (45.4)
Nitric oxide	10102439	Nitrogen oxide NO	1*	4	P076	A	10 (4.54)
p-Nitroaniline	100016	Benzenamine, 4-nitro-	1*	4	P077	D	5000 (2270)
Nitrobenzene	98953	Benzene, nitro-	1000	1,2,4	U169	C	1000 (454)
Nitrogen dioxide	10102440	Nitrogen oxide NO2	1000	1,4	P078	A	10 (4.54)
Nitrogen oxide NO	10102439	Nitric oxide	1*	4	P076	A	10 (4.54)
Nitrogen oxide NO2	10102440	Nitrogen dioxide	1000	1,4	P078	A	10 (4.54)
Nitroglycerine	55630	1,2,3-Propanetriol, trinitrate-	1*	4	P081	A	10 (4.54)
Nitrophenol (mixed)	25154556		1000	1		B	100 (45.4)
m-Nitrophenol	554847	2-Nitrophenol				B	100 (45.4)
o-Nitrophenol	88755	Phenol, 4-nitro-					
p-Nitrophenol	100027	4-Nitrophenol					**
o-Nitrophenol	88755	2-Nitrophenol	1000	1,2		B	100 (45.4)
p-Nitrophenol	100027	Phenol, 4-nitro- 4-Nitrophenol	1000	1,2,4		B	100 (45.4)
2-Nitrophenol	88755	o-Nitrophenol	1000	1,2	U170	B	100 (45.4)
4-Nitrophenol	100027	p-Nitrophenol Phenol, 4-nitro-	1000	1,2,4	U170	B	100 (45.4)
NITROPHENOLS	N.A.						
2-Nitropropane	79469	Propane, 2-nitro-	1*	2	U171	A	10 (4.54)
NITROSAMINES	N.A.						
N-Nitrosodi-n-butylamine	924163	1-Butanamine, N-butyl-N-nitroso-	1*	4	U172	X	10 (4.54)
N-Nitrosodiethanolamine	1116547	Ethanol, 2,2'-(nitrosoimino)bis-	1*	4	U173	X	1 (0.454)
N-Nitrosodiethylamine	55185	Ethanamine, N-ethyl-N-nitroso-	1*	4	U174	X	1 (0.454)
N-Nitrosodimethylamine	62759	Methanamine, N-methyl-N-nitroso-	1*	2,4	P082	A	10 (4.54)
N-Nitrosodiphenylamine	86306		1*	2		B	100 (45.4)
N-Nitroso-N-ethylurea	759739	Urea, N-ethyl-N-nitroso-	1*	4	U176	X	1 (0.454)
N-Nitroso-N-methylurea	684935	Urea, N-methyl-N-nitroso	1*	4	U177	X	1 (0.454)

Substance	CAS No.	Regulatory Synonyms	Statutory RQ	Code	RCRA Waste No.	Category	Final RQ lb (kg)
N-Nitroso-N-methylurethane	615532	Carbamic acid, methylnitroso-, ethyl ester	1*	4	U178	X	1 (0.454)
N-Nitrosomethylvinylamine	4549400	Vinylamine, N-methyl-N-nitroso-	1*	4	P084	A	10 (4.54)
N-Nitrosopiperidine	100754	Piperidine, 1-nitroso-	1*	4	U179	A	10 (4.54)
N-Nitrosopyrrolidine	930552	Pyrrolidine, 1-nitroso-	1*	4	U180	X	1 (0.454)
Nitrotoluene	1321126		1000	1		C	1000 (454)
m-Nitrotoluene	99081						
o-Nitrotoluene	88722						
p-Nitrotoluene	99990						
5-Nitro-o-toluidine	99558	Benzenamine, 2-methyl-5-nitro-	1*	4	U181	B	100 (45.4)
Octamethylpyrophosphoramide	152169	Diphosphoramide, octamethyl-	1*	4	P085	B	100 (45.4)
Osmium oxide OsO4 (T-4)-	20816120	Osmium tetroxide	1*	4	P087	C	1000 (454)
Osmium tetroxide	20816120	Osmium oxide OsO4 (T-4)-	1*	4	P087	C	1000 (454)
7-Oxabicyclo[2.2.1]heptane-2,3-dicarboxylic acid	145733	Endothall	1*	4	P088	C	1000 (454)
1,2-Oxathiolane, 2,2-dioxide	1120714	1,3-Propane sultone	1*	4	U193	A	10 (4.54)
2H-1,3,2-Oxazaphosphorin-2-amine, N,N-bis(2-chloroethyl)tetrahydro-, 2-oxide	50180	Cyclophosphamide	1*	4	U058	A	10 (4.54)
Oxirane	75218	Ethylene oxide	1*	4	U115	A	10 (4.54)
Oxiranecarboxyaldehyde	765344	Glycidylaldehyde	1*	4	U126	A	10 (4.54)
Oxirane, (chloromethyl)-	106898	Epichlorohydrin	1*	1,4	U041	B	100 (45.4)
Paraformaldehyde	30525894		1000	1		C	1000 (454)
Paraldehyde	123637	1,3,5-Trioxane, 2,4,6-trimethyl-	1000	4	U182	C	1000 (454)
Parathion	56382	Phosphorothioic acid, O,O-diethyl O-(4-nitrophenyl) ester.	1	1,4	P089	A	10 (4.54)
Pentachlorobenzene	608935	Benzene, pentachloro-	1*	4	U183	A	10 (4.54)
Pentachloroethane	76017	Ethane, pentachloro-	1*	4	U184	A	10 (4.54)
Pentachloronitrobenzene (PCNB)	82688	Benzene, pentachloronitro-	1*	4	U185	B	100 (45.4)
Pentachlorophenol	87865	Phenol, pentachloro-	10	1,2,4	U242	A	10 (4.54)
1,3-Pentadiene	504609	1-Methylbutadiene	1*	4	U186	B	100 (45.4)
Perchloroethylene	127184	Ethene, tetrachloro-; Tetrachloroethylene.	1*	2,4	U210	B	100 (45.4)
Phenacetin	62442	Acetamide, N-(4-ethoxyphenyl)-	1*	4	U187	B	100 (45.4)
Phenanthrene	85018		1*	2		D	5000 (2270)
Phenol	108952	Benzene, hydroxy-	1000	1,2,4	U188	C	1000 (454)
Phenol, 2-chloro-	95578	o-Chlorophenol; 2-Chlorophenol	1*	2,4	U048	B	100 (45.4)
Phenol, 4-chloro-3-methyl-	59507	p-Chloro-m-cresol; 4-Chloro-m-cresol	1*	2,4	U039	D	5000 (2270)
Phenol, 2-cyclohexyl-4,6-dinitro-	131895	2-Cyclohexyl-4,6-dinitrophenol	1*	4	P034	B	100 (45.4)
Phenol, 2,4-dichloro-	120832	2,4-Dichlorophenol	1*	2,4	U081	B	100 (45.4)
Phenol, 2,6-dichloro-	87650	2,6-Dichlorophenol	1*	4	U082	B	100 (45.4)
Phenol, 4,4'-(1,2-diethyl-1,2-ethenediyl)bis-, (E)	56531	Diethylstilbestrol	1*	4	U089	X	1 (0.454)
Phenol, 2,4-dimethyl-	105679	2,4-Dimethylphenol	1*	2,4	U101	B	100 (45.4)
Phenol, 2,4-dinitro-	51285	2,4-Dinitrophenol	1000	1,2,4	P048	A	10 (4.54)
Phenol, methyl-	1319773	Cresol(s) Cresylic acid	1000	1,4	U052	C	1000 (454)
m-Cresol	108394	m-Cresylic acid					
o-Cresol	95487	o-Cresylic acid					
p-Cresol	106445	p-Cresylic acid					
Phenol, 2-methyl-4,6-dinitro-	534521	4,6-Dinitro-o-cresol and salts	1*	2,4	P047	A	10 (4.54)
Phenol, 2,2'-methylenebis[3,4,6-trichloro-	70304	Hexachlorophene	1*	4	U132	B	100 (45.4)
Phenol, 2-(1-methylpropyl)-4,6-dinitro-	88857	Dinoseb	1*	4	P020	C	1000 (454)
Phenol, 4-nitro-	100027	p-Nitrophenol; 4-Nitrophenol	1000	1,2,4	U170	B	100 (45.4)

Hazardous substance	CASRN	Regulatory synonyms	Statutory			Final RQ	
			RQ	Code†	RCRA waste Number	Category	Pounds (Kg)
Phenol, pentachloro-	87865	Pentachlorophenol	10	1,2,4	U242	A	10 (4.54)
Phenol, 2,3,4,6-tetrachloro-	58902	2,3,4,6-Tetrachlorophenol	1*	4	U212	A	10 (4.54)
Phenol, 2,4,5-trichloro-	95954	2,4,5-Trichlorophenol	10	1,4	U230	A	10 (4.54)
Phenol, 2,4,6-trichloro-	88062	2,4,6-Trichlorophenol	10	1,2,4	U231	A	10 (4.54)
Phenol, 2,4,6-trinitro-, ammonium salt	131748	Ammonium picrate	1*	4	P009	A	10 (4.54)
L-Phenylalanine, 4-[bis(2-chloroethyl) aminol]	148823	Melphalan	1*		U150	X	1 (0.454)
1,10-(1,2-Phenylene)pyrene	193395	Indeno(1,2,3-cd)pyrene	1*	2,4	U137	B	100 (45.4)
Phenylmercury acetate	62384	Mercury, (acetato-O)phenyl-	1*	4	P092	B	100 (45.4)
Phenylthiourea	103855	Thiourea, phenyl-	1*	4	P093	B	100 (45.4)
Phorate	298022	Phosphorodithioic acid, O,O-diethyl S-(ethylthio), methyl ester.	1*	4	P094	A	10 (4.54)
Phosgene	75445	Carbonic dichloride	5000	1,4	P095	A	10 (4.54)
Phosphine	7803512		1*	4	P096	B	100 (45.4)
Phosphoric acid	7664382		5000	1		D	5000 (2270)
Phosphoric acid, diethyl 4-nitrophenyl ester	311455	Diethyl-p-nitrophenyl phosphate	1*	4	P041	B	100 (45.4)
Phosphoric acid, lead(2+) salt (2:3)	7446277	Lead phosphate	1*	4	U145	A	10 (4.54)
Phosphorodithioic acid, O,O-diethyl S-[2-(ethylthio)ethyl]ester	298044	Disulfoton	1	1,4	P039	X	1 (0.454)
Phosphorodithioic acid, O,O-diethyl S-(ethylthio), methyl ester	298022	Phorate	1*	4	P094	A	10 (4.54)
Phosphorodithioic acid, O,O-diethyl S-methyl ester	3288582	O,O-Diethyl S-methyl dithiophosphate	1*	4	U087	D	5000 (2270)
Phosphorodithioic acid, O,O-dimethyl S-[2(methylamino)-2-oxoethyl] ester	60515	Dimethoate	1*	4	P044	A	10 (4.54)
Phosphorofluoridic acid, bis(1-methylethyl) ester	55914	Diisopropylfluorophosphate	1*	4	P043	B	100 (45.4)
Phosphorothioic acid, O,O-diethyl O-(4-nitrophenyl) ester	56382	Parathion	1	1,4	P089	A	10 (4.54)
Phosphorothioic acid, O,[4-[(dimethylamino) sulfonyl]phenyl]O,O-dimethyl ester	52857	Famphur	1*	4	P097	C	1000 (454)
Phosphorothioic acid, O,O-dimethyl O-(4-nitrophenyl) ester	298000	Methyl parathion	100	1,4	P071	B	100 (45.4)
Phosphorothioic acid, O,O-diethyl O-pyrazinyl ester	297972	O,O-Diethyl O-pyrazinyl phosphorothioate	1*	4	P040	B	100 (45.4)
Phosphorus	7723140		1	1		X	1 (0.454)
Phosphrous oxycloride	10025873		5000	1		C	1000 (454)
Phosphorus pentasulfide	1314803	Phosphorus sulfide Sulfur phosphide	100	1,4	U189	B	100 (45.4)
Phosphorus sulfide	1314803	Phosphorus pentasulfide Sulfur phosphide	100	1,4	U189	B	100 (45.4)
Phosphorus trichloride	7719122		5000	1		C	1000 (454)**
PHTHALATE ESTERS	N.A.		1*	2			
Phthalic anhydride	85449	1,3-Isobenzofurandione	1*	4	U190	D	5000 (2270)
2-Picoline	109068	Pyridine, 2-methyl-	1*	4	U191	D	5000 (2270)
Piperidine, 1-nitroso-	100754	N-Nitrosopiperidine	1*	4	U179	A	10 (4.54)
Plumbane, tetraethyl-	78002	Tetraethyl lead	100	1,4	P110	A	10 (4.54)

Hazardous Substance	Regulatory Synonym	CAS No.	Stat. RQ	Code	RCRA	Category	Final RQ (lbs/kg)
POLYCHLORINATED BIPHENYLS (PCBs)	POLYCHLORINATED BIPHENYLS (PCBs)	1336363	10	2		A	1 (0.454)
Aroclor 1016	POLYCHLORINATED BIPHENYLS (PCBs)	12674112		1			----
Aroclor 1221	POLYCHLORINATED BIPHENYLS (PCBs)	11104282		1			----
Aroclor 1232	POLYCHLORINATED BIPHENYLS (PCBs)	11141165		1			----
Aroclor 1242	POLYCHLORINATED BIPHENYLS (PCBs)	53469219		1			----
Aroclor 1248	POLYCHLORINATED BIPHENYLS (PCBs)	12672296		1			----
Aroclor 1254	POLYCHLORINATED BIPHENYLS (PCBs)	11097691		1			----
Aroclor 1260	POLYCHLORINATED BIPHENYLS (PCBs)	11096825		1			----
POLYNUCLEAR AROMATIC HYDROCARBONS		N.A.	1*				**
Potassium arsenate	Potassium arsenate	7784410	1000			A	1 (0.454)
Potassium arsenite	Potassium arsenite	10124502	1000			A	1 (0.454)
Potassium bichromate	Potassium bichromate	7778509	1000			B	10 (4.54)
Potassium chromate	Potassium chromate	7789006	1000			B	10 (4.54)
Potassium cyanide K (CN)	Potassium cyanide K (CN)	151508	10	1,4	P098	B	10 (4.54)
Potassium cyanide K(CN)	Potassium cyanide	151508	10	1,4	P098	B	10 (4.54)
Potassium hydroxide		1310583	1000	1		D	1000 (454)
Potassium permanganate		7722647	100	1		C	100 (45.4)
Potassium silver cyanide	Argentate (1-), bis(cyano-C)-, potassium	506616	1*	4	P099	A	1 (0.454)
Pronamide	Benzamide, 3,5-dichloro-N-(1,1-dimethyl-2-propynyl)-	23950585	1*	4	U192	X	5000 (2270)
Propanal, 2-methyl-2-(methylthio)-, O-[(methylamino)carbonyl]oxime	Aldicarb	116063	1*	4	P070	A	1 (0.454)
1-Propanamine	n-Propylamine	107108	1*	4	U194	X	5000 (2270)
1-Propanamine, N-propyl-	Dipropylamine	142847	1*	4	U110	X	5000 (2270)
1-Propanamine, N-nitroso-N-propyl-	Di-n-propylnitrosamine	621647	1*	2,4	U111	A	1 (0.454)
Propane, 1,2-dibromo-3-chloro-	1,2-Dibromo-3-chloropropane	96128	1*	4	U066	A	1 (0.454)
Propane, 2-nitro-	2-Nitropropane	79469	1*	4	U171	B	10 (4.54)
1,3-Propane sultone	1,2-Oxathiolane, 2,2-dioxide	1120714	1*	4	U193	B	10 (4.54)
Propane, 1,2-dichloro-	Propylene dichloride; 1,2-Dichloropropane	78875	5000	1,2,4	U083	D	1000 (454)
Propanedinitrile	Malononitrile	109773	1*	4	U149	D	1000 (454)
Propanenitrile	Ethyl cyanide	107120	1*	4	P101	B	10 (4.54)
Propanenitrile, 3-chloro-	3-Chloropropionitrile	542767	1*	4	P027	D	1000 (454)
Propanenitrile, 2-hydroxy-2-methyl-	Acetone cyanohydrin; 2-Methyllactonitrile	75865	10	1,4	P069	B	10 (4.54)
Propane, 2,2'-oxybis[2-chloro-	Dichloroisopropyl ether	108601	1*	2,4	U027	D	1000 (454)
1,2,3-Propanetriol, trinitrate-	Nitroglycerine	55630	1*	4	P081	B	10 (4.54)
1-Propanol, 2,3-dibromo-, phosphate (3:1)	Tris(2,3-dibromopropyl) phosphate	126727	1*	4	U235	B	10 (4.54)
1-Propanol, 2-methyl-	Isobutyl alcohol	78831	1*	4	U140	X	5000 (2270)
2-Propanone	Acetone	67641	1*	4	U002	X	5000 (2270)
2-Propanone, 1-bromo-	Bromoacetone	598312	10	1	P017	D	1000 (454)
Propargite		2312358	1*			D	1000 (454)
Propargyl alcohol	2-Propyn-1-ol	107197	1*	1,2,4	P102	D	1000 (454)
2-Propenal	Acrolein	107028	1*	4	P003	A	1 (0.454)
2-Propenamide	Acrylamide	79061	1*	4	U007	X	5000 (2270)
1-Propene, 1,1,2,3,3,3-hexachloro-	Hexachloropropene	1888717	5000	4	U243	D	1000 (454)
1-Propene, 1,3-dichloro-	1,3-Dichloropropene	542756	100	1,2,4	U084	C	100 (45.4)
2-Propenenitrile	Acrylonitrile	107131	1*	1,2,4	U009	C	100 (45.4)
2-Propenenitrile, 2-methyl-	Methacrylonitrile	126987	1*	4	U152	D	1000 (454)
2-Propenoic acid	Acrylic acid	79107	1*	4	U008	X	5000 (2270)

Hazardous substance	Regulatory synonyms	CASRN	Statutory RQ	Statutory Code †	Statutory RCRA waste Number	Final RQ Category	Final RQ Pounds (Kg)
2-Propenoic acid, ethyl ester	Ethyl acrylate	140885	1*	4	U113	C	1000 (454)
2-Propenoic acid, 2-methyl-, ethyl ester	Ethyl methacrylate	97632	1*	4	U118	C	1000 (454)
2-Propenoic acid, 2-methyl-, methyl ester	Methyl methacrylate	80626	5000	1,4	U162	C	1000 (454)
2-Propen-1-ol	Allyl alcohol	107186	100	1,4	P005	B	100 (45.4)
Propionic acid		79094	5000	1		D	5000 (2270)
Propionic acid, 2-(2,4,5-trichlorophenoxy)-	Silvex (2,4,5-TP); 2,4,5-TP acid	93721	100	1,4	U233	B	100 (45.4)
Propionic anhydride		123626	5000	1		D	5000 (2270)
n-Propylamine	1-Propanamine	107108	5000	1	U194	D	5000 (2270)
Propylene dichloride	Propane, 1,2-dichloro-	78875	5000	1,2,4	U083	C	1000 (454)
Propylene oxide		75569	5000	1		B	100 (45.4)
1,2-Propylenimine	Aziridine, 2-methyl-	75558	1*	4	P067	X	1 (0.454)
2-Propyn-1-ol	Propargyl alcohol	107197	1*	4	P102	C	1000 (454)
Pyrene		129000	1*	2		D	5000 (2270)
Pyrethrins		121299 121211 8003347	1000	1		X	1 (0.545)
3,6-Pyridazinedione, 1,2-dihydro-	Maleic hydrazide	123331	1*	4	U148	D	5000 (2270)
4-Pyridinamine	4-Aminopyridine	504245	1*	4	P008	C	1000 (454)
Pyridine		110861	1*	4	U196	C	1000 (454)
2-Picoline	Pyridine, 2-methyl-	109068	1*	4	U191	D	5000 (2270)
Nicotine, & salts	Pyridine, 3-(1-methyl-2-pyrrolidinyl)-, (S)-	54115	1*	4	P075	B	100 (45.4)
Uracil mustard	2,4-(1H,3H)-Pyrimidinedione, 5-[bis(2-chloroethyl)amino]-	66751	1*	4	U237	A	10 (4.54)
Methylthiouracil	4(1H)-Pyrimidinone, 2,3-dihydro-6-methyl-2-thioxo-	56042	1*	4	U164	A	10 (4.54)
Pyrrolidine, 1-nitroso-	N-nitrosopyrrolidine	930552	1*	4	U180	X	1 (0.454)
Quinoline		91225	1000	1		D	5000 (2270)
RADIONUCLIDES		N.A.	1*	3			5000 (2270) §
Reserpine	Yohimban-16-carboxylic acid, 11,17-dimethoxy-18-[(3,4,5-trimethoxybenzoyl)oxy]-, methyl ester (3beta,16beta,17alpha,18beta,20alpha)-	50555	1*	4	U200	D	5000 (2270)
Resorcinol	1,3-Benzenediol	108463	1000	1,4	U201	D	5000 (2270)
Saccharin and salts	1,2-Benzisothiazol-3(2H)-one, 1,1-dioxide	81072	1*	4	U202	B	100 (45.4)
Safrole	1,3-Benzodioxole, 5-(2-propenyl)-	94597	1*	4	U203	B	100 (45.4)
Selenious acid		7783008	1*	4	U204	A	10 (4.54)
Selenious acid, dithallium (1+) salt	Thallium selenite	12039520	1*	4	P114	C	1000 (454)
Selenium ††		7782492	1*	2		B	100 (45.4) **
SELENIUM AND COMPOUNDS		N.A.		2			
Selenium dioxide	Selenium oxide	7446084	1000	1,4	U204	A	10 (4.54)
Selenium oxide	Selenium dioxide	7446084	1000	1,4	U204	A	10 (4.54)

Substance	CAS No.	Statutory RQ	Code	RCRA Waste No.	Category	Final RQ Pounds (Kg)
Selenium sulfide	7488564	1*	4	U205	A	10 (4.54)
Selenium sulfide SeS2	7488564	1*	4	U205	A	10 (4.54)
Selenourea	630104	1*	4	P103	C	1000 (454)
L-Serine, diazoacetate (ester)	115026	1*	4	U015	X	1 (0.454)
Silver††	7440224	1*	4		C	1000 (454)**
SILVER AND COMPOUNDS	N.A.		2			
Silver cyanide	506649	1*	2	P104	X	1 (0.454)
Silver cyanide Ag (CN)	506649	1*	4	P104	X	1 (0.454)
Silver nitrate	7761888	1	4		X	1 (0.454)
Silvex (2,4,5-TP)	93721	100	1	U233	B	100 (45.4)
Propionic acid, 2-(2,4,5-trichlorophenoxy)-			1,4			
2,4,5-TP acid						
Sodium	7440235	1000	1		A	10 (4.54)
Sodium arsenate	7631892	1000	1		X	1 (0.454)
Sodium arsenite	7784465	1000	1		X	1 (0.454)
Sodium azide	26628228	1*	4	P105	C	1000 (454)
Sodium bichromate	10588019	1000	1		A	10 (4.54)
Sodium bifluoride	1333831	5000	1		B	100 (45.4)
Sodium bisulfite	7631905	5000	1		D	5000 (2270)
Sodium chromate	7775113	1000	1		A	10 (4.54)
Sodium cyanide	143339	10	1,4	P106	A	10 (4.54)
Sodium cyanide Na (CN)	143339	10	1,4	P106	A	10 (4.54)
Sodium dodecylbenzenesulfonate	25155300	1000	1		C	1000 (454)
Sodium fluoride	7681494	5000	1		C	5000 (2270)
Sodium hydrosulfide	16721805	5000	1		D	5000 (2270)
Sodium hydroxide	1310732	1000	1		C	1000 (454)
Sodium hypochlorite	7681529	100	1		B	100 (45.4)
	10022705					
Sodium methylate	124414	1000	1		C	1000 (454)
Sodium nitrite	7632000	100	1		B	100 (45.4)
Sodium phosphate, dibasic	7558794	5000	1		D	5000 (2270)
	10039324					
	10140655					
Sodium phosphate, tribasic	7601549	5000	1		D	5000 (2270)
	7758294					
	7785844					
	101101890					
	10124568					
	10361894					
	10102188					
Sodium selenite	7782823	1000	1		B	100 (45.4)
Streptozotocin	18883664	1*	4	U206	X	1 (0.454)
D-Glucose, 2-deoxy-2-[[(methylnitrosoamino)-carbonyl]amino]-, 2-deoxy-2-(3-methyl-3-nitrosoureido)-Glucopyranose,						
Strontium chromate	7789062	1000	1		A	10 (4.54)
Strychnidin-10-one	57249	10	1,4	P108	A	10 (4.54)
Strychnidin-10-one, 2,3-dimethoxy-	357573	1*	1	P018	B	100 (45.4)
Strychnine, & salts	57249	10	1,4	P108	A	10 (4.54)
Styrene	100425	1000	1		C	1000 (454)
Sulfur monochloride	12771083	1000	1		C	1000 (454)

Hazardous substance	CASRN	Regulatory synonyms	Statutory			Final RQ	
			RQ	Code†	RCRA waste Number	Category	Pounds (Kg)
Sulfur phosphide	1314803	Phosphorus pentasulfide / Phosphorus sulfide	100	1,4	U189	B	100 (45.4)
Sulfuric acid	7664939 / 8014957		1000	1		C	1000 (454)
Sulfuric acid, dithallium (1+) salt	7446186 / 10031591	Thallium (I) sulfate	1000	1,4	P115	B	100 (45.4)
Sulfuric acid, dimethyl ester	77781	Dimethyl sulfate	1*	4	U103	B	100 (45.4)
2,4,5-T acid	93765	Acetic acid, (2,4,5-trichlorophenoxy) / 2,4,5-T	100	1,4	U232	C	1000 (454)
2,4,5-T amines	2008460 / 1319728 / 3813147 / 6369966 / 6369977		100	1		D	5000 (2270)
2,4,5-T esters	93798 / 1928478 / 2545597 / 25168154 / 61792072		100	1		C	1000 (454)
2,4,5-T salts	13560991		100	1		C	1000 (454)
2,4,5-T	93765	Acetic acid, (2,4,5-trichlorophenoxy) / 2,4,5-T acid	100	1,4	U232	C	1000 (454)
TDE	72548	Benzene, 1,1'-(2,2-dichloroethylidene)bis[4-chloro- / DDD 4,4' DDD	1	1,2,4	U060	X	1 (0.464)
1,2,4,5-Tetrachlorobenzene	95943	Benzene, 1,2,4,5-tetrachloro-	1*	4	U207	D	5000 (2270)
2,3,7,8-Tetrachlorodibenzo-p-dioxin (TCDD)	1746016		1*	2		X	1 (0.464)
1,1,1,2-Tetrachloroethane	630206	Ethane, 1,1,1,2-tetrachloro-	1*	2,4	U208	B	100 (45.4)
1,1,2,2-Tetrachloroethane	79345	Ethane, 1,1,2,2-tetrachloro-	1*	2,4	U209	B	100 (45.4)
Tetrachloroethene	127184	Ethene, tetrachloro- / Perchloroethylene / Tetrachloroethylene	1*	2,4	U210	B	100 (45.4)
Tetrachloroethylene	127184	Ethene, tetrachloro- / Perchloroethylene Tetrachloroethene	1*	2,4	U210	B	100 (45.4)
2,3,4,6-Tetrachlorophenol	58902	Phenol, 2,3,4,6-tetrachloro-	1*	4	U212	A	10 (4.54)
Tetraethyl lead	78002	Plumbane, tetraethyl-	100	1,4	P110	A	10 (4.54)
Tetraethyl pyrophosphate	107493	Diphosphoric acid, tetraethyl ester	100	1,4	P111	A	10 (4.54)
Tetraethyldithiopyrophosphate	3689245	Thiodiphosphoric acid, tetraethyl ester	1*	4	P109	B	100 (45.4)
Tetrahydrofuran	109999	Furan, tetrahydro-	1*	4	U213	C	1000 (454)
Tetranitromethane	509148	Methane, tetranitro-	1*	4	P112	A	10 (4.54)
Tetraphosphoric acid, hexaethyl ester	757584	Hexaethyl tetraphosphate	1*	4	P062	B	100 (45.4)
Thallic oxide	1314325	Thallium oxide Tl2O3	1*	4	P113	B	100 (45.4)

CASRN	Hazardous Substance	Regulatory Synonyms	Statutory RQ	Code	RCRA Waste No.	Final RQ Category	Final RQ Pounds (Kg)
7440280	Thallium ††		1*	2		C	1000 (454)
N.A.	Thallium and compounds		1*	2		B	100 (45.4)**
563688	Thallium (I) acetate	Acetic acid, thallium(1+) salt	1*	4	U214	B	100 (45.4)
6533739	Thallium (I) carbonate	Carbonic acid, dithallium(1+) salt	1*	4	U215	B	100 (45.4)
7791120	Thallium (I) chloride	Thallium chloride TlCl	1*	4	U216	B	100 (45.4)
7791120	Thallium chloride TlCl	Thallium(I) chloride TlCl	1*	4	U216	B	100 (45.4)
10102451	Thallium (I) nitrate	Nitric acid, thallium (1+) salt	1*	4	U217	B	100 (45.4)
1314325	Thallium oxide Tl2O3	Thallic oxide	1*	4	P113	B	100 (45.4)
12039520	Thallium selenite	Selenious acid, dithallium(1+) salt	1*	4	P114	C	1000 (454)
7446186	Thallium (I) sulfate	Sulfuric acid, dithallium(1+) salt	1000	1,4	P115	B	100 (45.4)
10031591							
62555	Thioacetamide	Ethanethioamide	1*	4	U218	A	10 (4.54)
3689245	Thiodiphosphoric acid, tetraethyl ester	Tetraethyldithiopyrophosphate	1*	4	P109	B	100 (45.4)
39196184	Thiofanox	2-Butanone, 3,3-dimethyl-1-(methylthio)-, O[(methylamino)carbonyl] oxime.	1*	4	P045	B	100 (45.4)
541537	Thioimidodicarbonic diamide [(H2N)C(S)] 2NH	Dithiobiuret	1*	4	P049	B	100 (45.4)
74931	Thiomethanol	Methanethiol / Methylmercaptan	100	1,4	U153	B	100 (45.4)
137268	Thioperoxydicarbonic diamide [(H2N)C(S)] 2S2, tetramethyl-	Thiram	1*	4	U244	A	10 (4.54)
108985	Thiophenol	Benzenethiol	1*	4	P014	B	100 (45.4)
79196	Thiosemicarbazide	Hydrazinecarbothioamide	1*	4	P116	B	100 (45.4)
62566	Thiourea	Thiourea	1*	4	U219	A	10 (4.54)
5344821	Thiourea, (2-chlorophenyl)-	1-(o-Chlorophenyl)thiourea	1*	4	P026	B	100 (45.4)
86884	Thiourea, 1-naphthalenyl-	alpha-Naphthylthiourea	1*	4	P072	B	100 (45.4)
103855	Thiourea, phenyl-	Phenylthiourea	1*	4	P093	B	100 (45.4)
137268	Thiram	Thioperoxydicarbonic diamide [(H2N)C(S)] 2S2, tetramethyl-	1*	4	U244	A	10 (4.54)
108883	Toluene	Benzene, methyl-	1000	1,2,4	U220	C	1000 (454)
95807, 496720, 823405, 25376458	Toluenediamine	Benzenediamine, ar-methyl-	1*	4	U221	A	10 (4.54)
584849, 91087, 26471625	Toluene diisocyanate	Benzene, 1,3-diisocyanatomethyl-	1*	4	U223	B	100 (45.4)
95534	o-Toluidine	Benzenamine, 2-methyl-	1*	4	U328	B	100 (45.4)
106490	p-Toluidine	Benzenamine, 4-methyl-	1*	4	U353	B	100 (45.4)
636215	o-Toluidine hydrochloride	Benzenamine, 2-methyl-, hydrochloride	1*	4	U222	B	100 (45.4)
8001352	Toxaphene	Camphene, octachloro-	100	1,2,4	P123	X	1 (0.454)
93721	2,4,5-TP acid	Propionic acid, 2-(2,4,5-trichlorophenoxy)- / Silvex (2,4,5-TP)	100	1,4	U233	B	100 (45.4)
32534955	2,4,5-TP esters		1*	1		B	100 (45.4)
61825	1H-1,2,4-Triazol-3-amine	Amitrole	1000	4	U011	A	10 (4.54)
52686	Trichlorfon		1*	1		B	100 (45.4)
120821	1,2,4-Trichlorobenzene		1*	2		B	100 (45.4)
71556	1,1,1-Trichloroethane	Ethane, 1,1,1-trichloro-	1*	2,4	U226	C	1000 (454)
	Methyl chloroform						
79005	1,1,2-Trichloroethane	Ethane, 1,1,2-trichloro-	1*	2,4	U227	B	100 (45.4)

Hazardous substance	CASRN	Statutory			Final RQ	
		RQ	Code†	RCRA waste Number	Category	Pounds (Kg)
Trichloroethene	79016	1000	1,2,4	U228	B	100 (45.4)
Trichloroethylene	79016	1000	1,2,4	U228	B	100 (45.4)
Trichloromethanesulfenyl chloride	594423	1*	4	P118	B	100 (45.4)
Trichloromonofluoromethane	75694	1*	4	U121	D	5000 (2270)
Trichlorophenol	25167822	10	1		A	10 (4.54)
2,3,4-Trichlorophenol	15950660					
2,3,5-Trichlorophenol	933788					
2,3,6-Trichlorophenol	933755					
2,4,5-Trichlorophenol	95954	10*	1,4	U230	A	10 (4.54)
2,4,6-Trichlorophenol	88062	10*	1,2,4	U231	A	10 (4.54)
3,4,5-Trichlorophenol	609198					
2,4,5-Trichlorophenol	95954	10*	1,4	U230	A	10 (4.54)
2,4,6-Trichlorophenol	88062	10	1,2,4	U231	A	10 (4.54)
Triethanolamine dodecylbenzenesulfonate	27323417	1000	1		C	1000 (454)
Triethylamine	121448	5000	1		D	5000 (2270)
Trimethylamine	75503	1000	1		B	100 (45.4)
1,3,5-Trinitrobenzene	99354	1*	4	U234	A	10 (4.54)
1,3,5-Trioxane, 2,4,6-trimethyl-	123637	1*	4	U182	C	1000 (454)
Tris(2,3-dibromopropyl) phosphate	126727	1*	4	U235	A	10 (4.54)
Trypan blue	72571	1*	4	U236	A	10 (4.54)
Unlisted Hazardous Wastes Characteristic of Corrosivity.	N.A.	1*	4	D002	B	100 (45.4)
Unlisted Hazardous Wastes Characteristics: Characteristic of Toxicity:	N.A.	1*	4			
Arsenic (D004)	N.A.	*1	4	D004	X	1 (0.454)
Barium (D005)	N.A.	*1	4	D005	C	1,000 (454)
Benzene (D018)	N.A.	1000	1, 2, 3, 4	D018	A	10 (4.54)
Cadmium (D006)	N.A.	*1	4	D006	A	10 (4.54)
Carbon tetrachloride (D019)	N.A.	5,000	1, 2, 4	D019	A	10 (4.54)
Chlordane (D020)	N.A.	1	1, 2, 4	D020	X	1 (0.454)
Chlorobenzene (D021)	N.A.	100	1, 2, 4	D021	B	100 (45.4)
Chloroform (D022)	N.A.	5,000	1, 2, 4	D022	A	10 (4.54)
Chromium (D007)	N.A.	*1	4	D007	A	10 (4.54)
o-Cresol (D023)	N.A.	1,000	1, 4	D023	C	1,000 (454)
m-Cresol (D024)	N.A.	1,000	1, 4	D024	C	1,000 (454)
p-Cresol (D025)	N.A.	1,000	1, 4	D025	C	1,000 (454)
Cresol (D026)	N.A.	1,000	1, 4	D026	C	1,000 (454)

Regulatory synonyms:
- Trichloroethene: Ethene, trichloro-
- Trichloroethylene: Trichloroethylene / Ethene, trichloro-
- Trichloromethanesulfenyl chloride: Methanesulfenyl chloride, trichloro-
- Trichloromonofluoromethane: Methane, trichlorofluoro-
- 2,4,5-Trichlorophenol: Phenol, 2,4,5-trichloro-
- 2,4,6-Trichlorophenol: Phenol, 2,4,6-trichloro-
- 1,3,5-Trinitrobenzene: Benzene, 1,3,5-trinitro-
- 1,3,5-Trioxane, 2,4,6-trimethyl-: Paraldehyde
- Tris(2,3-dibromopropyl) phosphate: 1-Propanol, 2,3-dibromo-, phosphate [(3:1)
- Trypan blue: 2,7-Naphthalenedisulfonic acid, 3,3'-3,3'-dimethyl-(1,1'-biphenyl)-4,4'-diyl)-bis(azo)]bis(5-amino-4-hydroxy)-tetrasodium salt.

Hazardous Substance	CAS No.	Description	Statutory RQ	Code	RCRA Waste No.	Cat.	Final RQ
2,4-D (D016)	N.A.		100	1,4	D016	B	100 (45.4)
1,4-Dichlorobenzene (D027)	N.A.		100	1,2,4	D027	B	100 (45.4)
1,2-Dichloroethane (D028)	N.A.		5,000	1,2,4	D028	B	100 (45.4)
1,1-Dichloroethylene (D029)	N.A.		5,000	1,2,4	D029	B	100 (45.4)
2,4-Dinitrotoluene (D030)	N.A.		1,000	1,2,4	D030	A	10 (4.54)
Endrin (D012)	N.A.		1*	4	D012	X	1 (0.454)
Heptachlor (and epoxide) (D031)	N.A.		1*	1,2,4	D031	X	1 (0.454)
Hexachlorobenzene (D032)	N.A.		1*	2,4	D032	A	10 (4.54)
Hexachlorobutadiene (D033)	N.A.		1*	2,4	D033	X	1 (0.454)
Hexachloroethane (D034)	N.A.		1*	2,4	D034	B	100 (45.4)
Lead (D008)	N.A.		1*	4	D008	A	10 (4.54)
Lindane (D013)	N.A.		1*	4	D013	X	1 (0.454)
Mercury (D009)	N.A.		1*	4	D009	X	1 (0.454)
Methoxychlor (D014)	N.A.		1*	4	D014	X	1 (0.454)
Methyl ethyl ketone (D035)	N.A.		1*	1,4	D035	D	5,000 (2270)
Nitrobenzene (D036)	N.A.		1,000	1,2,4	D036	C	1,000 (454)
Pentachlorophenol (D037)	N.A.		10	1,2,4	D037	C	10 (4.54)
Pyridine (D038)	N.A.		1*	4	D038	C	1,000 (454)
Selenium (D010)	N.A.		1*	4	D010	X	10 (4.54)
Silver (D011)	N.A.		1*	4	D011	B	1 (0.454)
Tetrachloroethylene (D039)	N.A.		1*	2,4	D039	X	100 (45.4)
Toxaphene (D015)	N.A.		1*	4	D015	B	1 (0.454)
Trichloroethylene (D040)	N.A.		1000	1,2,4	D040	B	100 (45.4)
2,4,5-Trichlorophenol (D041)	N.A.		10	1,2,4	D041	A	10 (4.54)
2,4,6-Trichlorophenol (D042)	N.A.		10	1,2,4	D042	A	10 (4.54)
2,4,5-TP (D017)	N.A.		100	1,4	D017	B	100 (45.4)
Vinyl chloride (D043)	N.A.		1*	2,3,4	D043	X	1 (0.454)
Unlisted Hazardous Wastes Characteristic of Ignitability.	N.A.				D001	B	100 (45.4)
Unlisted Hazardous Wastes Characteristic of Reactivity.	N.A.		1*	4	D003	B	100 (45.4)
Uracil mustard	66751	2,4-(1H,3H)-Pyrimidinedione, 5-[bis(2-chloroethyl)amino]-.	1*	4	U237	A	10 (4.54)
Uranyl acetate	541093		5000	1		B	100 (45.4)
Uranyl nitrate	10102064 / 36478769		5000	1		B	100 (45.4)
Urea, N-ethyl-N-nitroso-	759739	N-Nitroso-N-ethylurea	1*	4	U176	X	1 (0.454)
Urea, N-methyl-N-nitroso	684935	N-Nitroso-N-methylurea	1*	4	U177	X	1 (0.454)
Vanadic acid, ammonium salt	7803556	Ammonium vanadate	1*	4	P119	C	1000 (454)
Vanadium pentoxide	1314621	Vanadium oxide V205	1000	1,4	P120	C	1000 (454)
Vanadium oxide V205	1314621	Vanadium oxide V205	1000	1,4	P120	C	1000 (454)
Vanadyl sulfate	27774136		1*	1		X	1 (0.454)
Vinyl chloride	75014	Ethene, chloro-	1000	2,3,4	U043	D	5000 (2270)
Vinyl acetate	108054	Vinyl acetate monomer	1000	1		D	5000 (2270)
Vinyl acetate monomer	108054	Vinyl acetate	1*	1		A	10 (4.54)
Vinylamine, N-methyl-N-nitroso-	4549400	N-Nitrosomethylvinylamine	1*	1,2,4	P084	B	100 (45.4)
Vinylidene chloride	75354	Ethene, 1,1-dichloro- / 1,1-Dichloroethylene	5000		U078	B	100 (45.4)
Warfarin, & salts, when present at concentrations greater than 0.3%.	81812	2H-1-Benzopyran-2-one, 4-hydroxy-3-(3-oxo-1-phenyl-butyl)-, & salts, when present at concentrations greater than 0.3%.	1*	4	P001	B	100 (45.4)

Hazardous substance	CASRN	Regulatory synonyms	Statutory RQ	Statutory Code †	Statutory RCRA waste Number	Final RQ Category	Final RQ Pounds (Kg)
Xylene (mixed)	1330207	Benzene, dimethyl	1000	1,4	U239	C	1000 (454)
m-Benzene, dimethyl	108383	m-Xylene					
o-Benzene, dimethyl	95476	o-Xylene					
p-Benzene, dimethyl	106423	p-Xylene					
Xylenol	1300716		1000	1		C	1000 (454)
Yohimban-16-carboxylic acid,11,17-dimethoxy-18-[(3,4,5-trimethoxybenzoyl)oxy]-, methyl ester (3beta,16beta,17alpha,18beta, 20alpha)-	50555	Reserpine	1*	4	U200	D	5000 (2270)
Zinc††	7440666		1*	2		C	1000 (454)
ZINC AND COMPOUNDS	N.A.		1*	2			**
Zinc acetate	557346		1000	1		C	1000 (454)
Zinc ammonium chloride	52628258 14639975 14639986		5000	1		C	1000 (454)
Zinc borate	1332076		1000	1		C	1000 (454)
Zinc bromide	7699458		5000	1		C	1000 (454)
Zinc carbonate	3486359		1000	1		C	1000 (454)
Zinc chloride	7646857		5000	1		C	1000 (454)
Zinc cyanide	557211	Zinc cyanide Zn(CN)2	10	1,4	P121	A	10 (4.54)
Zinc cyanide Zn(CN)2	557211	Zinc cyanide	10	1,4	P121	A	10 (4.54)
Zinc fluoride	7783495		1000	1		C	1000 (454)
Zinc formate	557415		1000	1		C	1000 (454)
Zinc hydrosulfite	7779864		1000	1		C	1000 (454)
Zinc nitrate	7779886		5000	1		C	1000 (454)
Zinc phenosulfonate	127822		5000	1		D	5000 (2270)
Zinc phosphide	1314847	Zinc phosphide Zn3P2, when present at concentrations greater than 10%.	1000	1,4	P122	B	100 (45.4)
Zinc phosphide Zn3P2, when present at concentrations greater than 10%.	1314847	Zinc phosphide	1000	1,4	P122	B	100 (45.4)
Zinc silicofluoride	16871719		5000	1		D	5000 (2270)
Zinc sulfate	7733020		1000	1		D	1000 (454)
Zirconium nitrate	13746899		5000	1		D	5000 (2270)
Zirconium potassium fluoride	16923958		5000	1		C	1000 (454)
Zirconium sulfate	14644612		5000	1		D	5000 (2270)
Zirconium tetrachloride	10026116		5000	1		D	5000 (2270)
F001			1*	4	F001	A	10 (4.54)

The following spent halogenated solvents used in degreasing; all spent solvent mixtures/blends used in degreasing containing, before use, a total of ten percent or more (by volume) of one or more of the above halogenated solvents or those solvents listed in F002, F004, and F005; and still bottoms from the recovery of these spent solvents and spent solvent mixtures.

CAS No.	Hazardous Substance	Statutory RQ	Statutory Code	RCRA Waste No.	Final RQ Category	Final RQ Pounds (Kg)
127184	(a) Tetrachloroethylene	1*	2,4	U210	B	100 (45.4)
79016	(b) Trichloroethylene	1000	1,2,4	U228	B	100 (45.4)
75092	(c) Methylene chloride	1*	2,4	U080	C	1000 (454)
71556	(d) 1,1,1-Trichloroethane	1*	2,4	U226	C	1000 (454)
56235	(e) Carbon tetrachloride	5000	1,2,4	U211	A	10 (4.54)
N.A.	(f) Chlorinated fluorocarbons				D	5000 (2270)
	F002	1*	4	F002	A	10 (4.54)

The following spent halogenated solvents; all spent solvent mixtures/blends containing, before use, a total of ten percent or more (by volume) of one or more of the above halogenated solvents or those listed in F001, F004, or F005; and still bottoms from the recovery of these spent solvents and spent solvent mixtures.

CAS No.	Hazardous Substance	Statutory RQ	Statutory Code	RCRA Waste No.	Final RQ Category	Final RQ Pounds (Kg)
127184	(a) Tetrachloroethylene	1*	2,4	U210	B	100 (45.4)
75092	(b) Methylene chloride	1*	2,4	U080	C	1000 (454)
79016	(c) Trichloroethylene	1000	1,2,4	U228	B	100 (45.4)
71556	(d) 1,1,1-Trichloroethane	1*	2,4	U226	C	1000 (454)
108907	(e) Chlorobenzene	100	1,2,4	U037	B	100 (45.4)
76131	(f) 1,1,2-Trichloro-1,2,2-trifluoroethane				D	5000 (2270)
95501	(g) o-Dichlorobenzene	100	1,2,4	U070	B	100 (45.4)
75694	(h) Trichlorofluoromethane	1*	4	U121	D	5000 (2270)
79005	(i) 1,1,2-Trichloroethane	1*	2,4	U227	B	100 (45.4)
	F003	1*	4	F003	B	100 (45.4)

The following spent non-halogenated solvents and the still bottoms from the recovery of these solvents:

CAS No.	Hazardous Substance	Statutory RQ	Statutory Code	RCRA Waste No.	Final RQ Category	Final RQ Pounds (Kg)
1330207	(a) Xylene				C	1000 (454)
67641	(b) Acetone				D	5000 (2270)
141786	(c) Ethyl acetate				D	5000 (2270)
100414	(d) Ethylbenzene				C	1000 (454)
60297	(e) Ethyl ether				B	100 (45.4)
108101	(f) Methyl isobutyl ketone				D	5000 (2270)
71363	(g) n-Butyl alcohol				D	5000 (2270)
108941	(h) Cyclohexanone				D	5000 (2270)
67561	(i) Methanol				D	5000 (2270)
	F004	1*	4	F004	C	1000 (454)

The following spent non-halogenated solvents and the still bottoms from the recovery of these solvents:

CAS No.	Hazardous Substance	Statutory RQ	Statutory Code	RCRA Waste No.	Final RQ Category	Final RQ Pounds (Kg)
1319773	(a) Cresols/Cresylic acid	1000	1,4	U052	C	1000 (454)
98953	(b) Nitrobenzene	1000	1,2,4	U169	C	1000 (454)

Hazardous substance	CASRN	Statutory			Final RQ	
		RQ	Code †	RCRA waste Number	Category	Pounds (Kg)
F005 The following spent non-halogenated solvents and the still bottoms from the recovery of these solvents:		1°	4	F005	B	100 (45.4)
(a) Toluene	108883	1000	1,2,4	U220	C	1000 (454)
(b) Methyl ethyl ketone	78933	1°	4	U159	D	5000 (2270)
(c) Carbon disulfide	75150	5000	1,4	P022	B	100 (45.4)
(d) Isobutanol	78831	1°	4	U140	D	5000 (2270)
(e) Pyridine	110861	1°	4	U196	C	1000 (454)
F006 Wastewater treatment sludges from electroplating operations except from the following processes: (1) sulfuric acid anodizing of aluminum, (2) tin plating on carbon steel, (3) zinc plating (segregated basis) on carbon steel, (4) aluminum or zinc-aluminum plating on carbon steel, (5) cleaning/stripping associated with tin, zinc and aluminum plating on carbon steel, and (6) chemical etching and milling of aluminum.		1°	4	F006	A	10 (4.54)
F007 Spent cyanide plating bath solutions from electroplating operations.		1°	4	F007	A	10 (4.54)
F008 Plating bath residues from the bottom of plating baths from electroplating operations where cyanides are used in the process.		1°	4	F008	A	10 (4.54)
F009 Spent stripping and cleaning bath solutions from electroplating operations where cyanides are used in the process.		1°	4	F009	A	10 (4.54)
F010 Quenching bath residues from oil baths from metal heat treating operations where cyanides are used in the process.		1°	4	F010	A	10 (4.54)
F011 Spent cyanide solution from salt bath pot cleaning from metal heat treating operations.		1°	4	F011	A	10 (4.54)
F012		1°	4	F012	A	10 (4.54)

Substance			Statutory Code	RQ
F019 Quenching wastewater treatment sludges from metal heat treating operations where cyanides are used in the process.	1	4	A	10 (4.54)
F020 Wastes (except wastewater and spent carbon from hydrogen chloride purification) from the production or manufacturing use (as a reactant, chemical intermediate, or component in a formulating process) of tri-or-tetrachlorophenol, or of intermediates used to produce their pesticide derivatives. (This listing does not include wastes from the production of hexachlorophene from highly purified 2,4,5-trichlorophenol.)	1*	4	X	1 (0.454)
F021 Wastes (except wastewater and spent carbon from hydrogen chloride purification) from the production or manufacturing use (as a reactant, chemical intermediate, or component in a formulating process) of pentachlorophenol, or of intermediates used to produce its derivatives.	1*	4	X	1 (0.454)
F022 Wastes (except wastewater and spent carbon from hydrogen chloride purification) from the manufacturing use (as a reactant, chemical intermediate, or component in a formulating process) of tetra-, penta-, or hexachlorobenzenes under alkaline conditions.	1*	4	X	1 (0.454)
F023 Wastes (except wastewater and spent carbon from hydrogen chloride purification) from the production of materials on equipment previously used for the production or manufacturing use (as a reactant, chemical intermediate, or component in a formulating process) of tri- and tetrachlorophenols. (This listing does not include wastes from equipment used only for the production or use of hexachlorophene from highly purified 2,4,5-trichlorophenol.)	1*	4	X	1 (0.454)
F024	1*	4	X	1 (0.454)

Hazardous substance	CASRN	Regulatory synonyms	Statutory			Final RQ	
			RQ	Code†	RCRA waste Number	Category	Pounds (Kg)
Wastes, including but not limited to distillation residues, heavy ends, tars, and reactor cleanout wastes, from the production of chlorinated aliphatic hydrocarbons, having carbon content from one to five, utilizing free radical catalyzed processes. (This listing does not include light ends, spent filters and filter aids, spent dessicants(sic), wastewater, wastewater treatment sludges, spent catalysts, and wastes listed in Section 261.32.)							
F025 Condensed light ends, spent filters and filter aids, and spent desiccant wastes from the production of certain chlorinated aliphatic hydrocarbons, by free radical catalyzed processes. These chlorinated aliphatic hydrocarbons are those having carbon chain lengths ranging from one to and including five, with varying amounts and positions of chlorine substitution..			1*	4	F025	X	##1 (0.454)
F026 Wastes (except wastewater and spent carbon from hydrogen chloride purification) from the production of materials on equipment previously used for the manufacturing use (as a reactant, chemical intermediate, or component in a formulating process) of tetra-, penta-, or hexachlorobenzene under alkaline conditions.			1*	4	F026	X	1 (0.454)
F027 Discarded unused formulations containing tri-, tetra-, or pentachlorophenol or discarded unused formulations containing compounds derived from these chlorophenols. (This listing does not include formulations containing hexachlorophene synthesized from prepurified 2,4,5-tri-chlorophenol as the sole component.)			1*	4	F027	X	1 (0.454)
F028 Residues resulting from the incineration or thermal treatment of soil contaminated with EPA Hazardous Waste Nos. F020, F021, F022, F023, F026, and F027.			1*	4	F028	X	1 (0.454)
F032			1*	4	F032	X	1(0.454)

					Statutory			Final RQ

Description								RQ
Wastewaters (except those that have not come into contact with process contaminants), process residuals, preservative drippage, and spent formulations from wood preserving processes generated at plants that currently use or have previously used chlorophenolic formulations (except potentially cross-contaminated wastes that have had the F032 waste code deleted in accordance with §261.35 of this chapter or potentially cross-contaminated wastes that are otherwise currently regulated as hazardous wastes (i.e., F034 or F035), and where the generator does not resume or initiate use of chlorophenolic formulations). This listing does not include K001 bottom sediment sludge from the treatment of wastewater from wood preserving processes that use creosote and/or pentachlorophenol.	1*	4	F034	X				1(0.454)
F034								
Wastewaters (except those that have not come into contact with process contaminants), process residuals, preservative drippage, and spent formulations from wood preserving processes generated at plants that use creosote formulations. This listing does not include K001 bottom sediment sludge from the treatment of wastewater from wood preserving processes that use creosote and/or pentachlorophenol.	1*	4	F035	X				1(0.454)
F035								
Wastewaters (except those that have not come into contact with process contaminants), process residuals, preservative drippage, and spent formulations from wood preserving processes generated at plants that use inorganic preservatives containing arsenic or chromium. This listing does not include K001 bottom sediment sludge from the treatment of wastewater from wood preserving processes that use creosote and/or pentachlorophenol.	1*	4	F037	X				1 (0.454)
F037								

| Hazardous substance | CASRN | Regulatory synonyms | Statutory | | | Final RQ | |
			RQ	Code†	RCRA waste Number	Category	Pounds (Kg)
Petroleum refinery primary oil/water/solids separation sludge—Any sludge generated from the gravitational separation of oil/water/solids during the storage or treatment of process wastewaters and oily cooling wastewaters from petroleum refineries. Such sludges include, but are not limited to, those generated in: oil/water/solids separators; tanks and impoundments; ditches and other conveyances; sumps; and stormwater units receiving dry weather flow. Sludge generated in stormwater units that do not receive dry weather flow, sludges generated from non-contact once-through cooling waters segregated for treatment from other process or oily cooling waters, sludges generated in aggressive biological treatment units as defined in §261.31(b)(2) (including sludges generated in one or more additional units after wastewaters have been treated in aggressive biological treatment units) and K051 wastes are not included in this listing.			1*	4	F038	X	1 (0.454)
F038							

Code	Description				Statutory RQ
K001	Petroleum refinery secondary (emulsified) oil/water/solids separation sludge—Any sludge and/or float generated from the physical and/or chemical separation of oil/water/solids in process wastewaters and oily cooling wastewaters from petroleum refineries. Such wastes include, but are not limited to, all sludges and floats generated in: induced air flotation (IAF) units, tanks and impoundments, and all sludges generated in DAF units. Sludges generated in stormwater units that do not receive dry weather flow, sludges generated from once-through non-contact cooling waters, sludges and floats generated in aggressive biological treatment units as defined in §261.31(b)(2) (including sludges and floats generated in one or more additional units after wastewaters have been treated in aggressive biological treatment units) and F037, K048, and K051 wastes are not included in this listing.	x	1*	4	1 (0.454)
K002	Bottom sediment sludge from the treatment of wastewaters from wood preserving processes that use creosote and/or pentachlorophenol.	▲	1*	4	10 (4.54)
K003	Wastewater treatment sludge from the production of chrome yellow and orange pigments.	▲	1*	4	10 (4.54)
K004	Wastewater treatment sludge from the production of molybdate orange pigments.	▲	1*	4	10 (4.54)
K005	Wastewater treatment sludge from the production of zinc yellow pigments.	▲	1*	4	10 (4.54)
K006	Wastewater treatment sludge from the production of chrome green pigments.	▲	1*	4	10 (4.54)
K007	Wastewater treatment sludge from the production of chrome oxide green pigments (anhydrous and hydrated).	▲	1*	4	10 (4.54)
K008	Wastewater treatment sludge from the production of iron blue pigments.	▲	1*	4	10 (4.54)

Hazardous substance	CASRN	Regulatory synonyms	Statutory RQ	Statutory Code†	Statutory RCRA waste Number	Final RQ Category	Final RQ Pounds (Kg)
Oven residue from the production of chrome oxide green pigments. K009			1*	4	K009	A	10 (4.54)
Distillation bottoms from the production of acetaldehyde from ethylene. K010			1*	4	K010	A	10 (4.54)
Distillation side cuts from the production of acetaldehyde from ethylene. K011			1*	4	K011	A	10 (4.54)
Bottom stream from the wastewater stripper in the production of acrylonitrile. K013			1*	4	K013	A	10 (4.54)
Bottom stream from the acetonitrile column in the production of acrylonitrile. K014			1*	4	K014	D	5000 (2270)
Bottoms from the acetonitrile purification column in the production of acrylonitrile. K015			1*	4	K015	A	10 (4.54)
Still bottoms from the distillation of benzyl chloride. K016			1*	4	K016	X	1 (0.454)
Heavy ends or distillation residues from the production of carbon tetrachloride. K017			1*	4	K017	A	10 (4.54)
Heavy ends (still bottoms) from the purification column in the production of epi-chlorohydrin. K018			1*	4	K018	X	1 (0.454)
Heavy ends from the fractionation column in ethyl chloride production. K019			1*	4	K019	X	1 (0.454)
Heavy ends from the distillation of ethylene dichloride in ethylene dichloride production. K020			1*	4	K020	X	1 (0.454)
Heavy ends from the distillation of vinyl chloride in vinyl chloride monomer production. K021			1*	4	K021	A	10 (4.54)

Code	Substance			Category	RQ
K022	Aqueous spent antimony catalyst waste from fluoromethanes production.	4	1°	X	1 (0.454)
K023	Distillation bottom tars from the production of phenol/acetone from cumene.	4	1°	D	5000 (2270)
K024	Distillation light ends from the production of phthalic anhydride from naphthalene.	4	1°	D	5000 (2270)
K025	Distillation bottoms from the production of phthalic anhydride from naphthalene.	4	1°	A	10 (4.54)
K026	Distillation bottoms from the production of nitrobenzene by the nitration of benzene.	4	1°	C	1000 (454)
K027	Stripping still tails from the production of methyl ethyl pyridines.	4	1°	A	10 (4.54)
K028	Centrifuge and distillation residues from toluene diisocyanate production.	4	1°	X	1 (0.454)
K029	Spent catalyst from the hydrochlorinator reactor in the production of 1,1,1-trichloroethane.	4	1°	X	1 (0.454)
K030	Waste from the product steam stripper in the production of 1,1,1-trichloroethane.	4	1°	X	1 (0.454)
K031	Column bottoms or heavy ends from the combined production of trichloroethylene and perchloroethylene.	4	1°	X	1 (0.454)
K032	By-product salts generated in the production of MSMA and cacodylic acid.	4	1°	A	10 (4.54)
K033	Wastewater treatment sludge from the production of chlordane.	4	1°	A	10 (4.54)
K034	Wastewater and scrub water from the chlorination of cyclopentadiene in the production of chlordane.	4	1°	A	10 (4.54)

Hazardous substance	CASRN	Statutory			Final RQ	
		RQ	Code †	RCRA waste Number	Category	Pounds (Kg)
Filter solids from the filtration of hexachlorocyclopentadiene in the production of chlordane.						
K035		1*	4	K035	X	1 (0.454)
Wastewater treatment sludges generated in the production of creosote.						
K036		1*	4	K036	X	1 (0.454)
Still bottoms from toluene reclamation distillation in the production of disulfoton.						
K037		1*	4	K037	X	1 (0.454)
Wastewater treatment sludges from the production of disulfoton.						
K038		1*	4	K038	A	10 (4.54)
Wastewater from the washing and stripping of phorate production.						
K039		1*	4	K039	A	10 (4.54)
Filter cake from the filtration of diethylphosphorodithioic acid in the production of phorate.						
K040		1*	4	K040	A	10 (4.54)
Wastewater treatment sludge from the production of phorate.						
K041		1*	4	K041	X	1 (0.454)
Wastewater treatment sludge from the production of toxaphene.						
K042		1*	4	K042	A	10 (4.54)
Heavy ends or distillation residues from the distillation of tetrachlorobenzene in the production of 2,4,5-T.						
K043		1*	4	K043	A	10 (4.54)
2,6-Dichlorophenol waste from the production of 2,4-D.						
K044		1*	4	K044	A	10 (4.54)

Code	Substance				RQ
K045	Wastewater treatment sludges from the manufacturing and processing of explosives.	4	A	1*	10 (4.54)
K046	Spent carbon from the treatment of wastewater containing explosives.	4	A	1*	10 (4.54)
K047	Wastewater treatment sludges from the manufacturing, formulation and loading of lead-based initiating compounds.	4	A	1*	10 (4.54)
K048	Pink/red water from TNT operations.	4	A	1*	10 (4.54)
K049	Dissolved air flotation (DAF) float from the petroleum refining industry.	4	A	1*	10 (4.54)
K050	Slop oil emulsion solids from the petroleum refining industry.	4	A	1*	10 (4.54)
K051	Heat exchanger bundle cleaning sludge from the petroleum refining industry.	4	A	1*	10 (4.54)
K052	API separator sludge from the petroleum refining industry.	4	A	1*	10 (4.54)
K060	Ammonia still lime sludge from coking operations.	4	X	1*	1 (0.454)
K061	Emission control dust/sludge from the primary production of steel in electric furnaces.	4	A	1*	10 (4.54)
K062	Spent pickle liquor generated by steel finishing operations of facilities within the iron and steel industry (SIC Codes 331 and 332).	4	A	1*	10 (4.54)
K064	Acid plant blowdown slurry/sludge resulting from thickening of blowdown slurry from primary copper production.	4	A	1*	10 (4.54)
K065		4	A	1*	10 (4.54)

Hazardous substance	CASRN	Statutory			Final RQ	
		RQ	Code†	RCRA waste Number	Category	Pounds (Kg)
Surface impoundment solids contained in and dredged from surface impoundments at primary lead smelting facilities. K066		1*	4	K066	A	10 (4.54)
Sludge from treatment of process wastewater and/or acid plant blowdown from primary zinc production. K069		1*	4	K069	A	10 (4.54)
Emission control dust/sludge from secondary lead smelting. K071		1*	4	K071	X	1 (0.454)
Brine purification muds from the mercury cell process in chlorine production, where separately prepurified brine is not used. K073		1*	4	K073	A	10 (4.54)
Chlorinated hydrocarbon waste from the purification step of the diaphragm cell process using graphite anodes in chlorine production. K083		1*	4	K083	B	100 (45.4)
Distillation bottoms from aniline extraction. K084		1*	4	K084	X	1 (0.454)
Wastewater treatment sludges generated during the production of veterinary pharmaceuticals from arsenic or organo-arsenic compounds. K085		1*	4	K085	A	10 (4.54)
Distillation or fractionation column bottoms from the production of chlorobenzenes. K086		1*	4	K086	A	10 (4.54)
Solvent washes and sludges, caustic washes and sludges, or water washes and sludges from cleaning tubs and equipment used in the formulation of ink from pigments, driers, soaps, and stabilizers containing chromium and lead. K087		1*	4	K087	B	100 (45.4)
Decanter tank tar sludge from coking operations. K088		1*	4	K088		

Code	Description				RQ		Category
K090	Spent potliners from primary aluminum reduction.	4	1*				
K091	Emission control dust or sludge from ferrochromiumsilicon production.	4	1				
K093	Emission control dust or sludge from ferrochromium production.	4	1*		5000 (2270)		D
K094	Distillation light ends from the production of phthalic anhydride from ortho-xylene.	4	1*		5000 (2270)		D
K095	Distillation bottoms from the production of phthalic anhydride from ortho-xylene.	4	1*		100 (45.4)		B
K096	Distillation bottoms from the production of 1,1,1-trichloroethane.	4	1*		100 (45.4)		B
K097	Heavy ends from the heavy ends column from the production of 1,1,1-trichloroethane.	4	1*		1 (0.454)		X
K098	Vacuum stripper discharge from the chlordane chlorinator in the production of chlordane.	4	1*		1 (0.454)		X
K099	Untreated process wastewater from the production of toxaphene.	4	1*		10 (4.54)		A
K100	Untreated wastewater from the production of 2,4-D.	4	1*		10 (4.54)		A
K101	Waste leaching solution from acid leaching of emission control dust/sludge from secondary lead smelting.	4	1*		1 (0.454)		X
K102	Distillation tar residues from the distillation of aniline-based compounds in the production of veterinary pharmaceuticals from arsenic or organo-arsenic compounds.	4	1*		1 (0.454)		X
K103	Residue from the use of activated carbon for decolorization in the production of veterinary pharmaceuticals from arsenic or organo-arsenic compounds.	4	1*		100 (45.4)		B

Hazardous substance	CASRN	Statutory			Final RQ	
		RQ	Code †	RCRA waste Number	Category	Pounds (Kg)
Process residues from aniline extraction from the production of aniline. K104	....	1*	4	K104	A	10 (4.54)
Combined wastewater streams generated from nitrobenzene/aniline production. K105	....	1*	4	K105	A	10 (4.54)
Separated aqueous stream from the reactor product washing step in the production of chlorobenzenes. K106	....	1*	4	K106	X	1 (0.454)
Wastewater treatment sludge from the mercury cell process in chlorine production. K107	....	10	4	K107	X	10 (4.54)
Column bottoms from product separation from the production of 1,1-dimethylhydrazine (UDMH) from carboxylic acid hydrazines. K108	....	10	4	K108	X	10 (4.54)
Condensed column overheads from product separation and condensed reactor vent gases from the production of 1,1-dimethylhydrazine (UDMH) from carboxylic acid hydrazides. K109	....	10	4	K109	X	10 (4.54)
Spent filter cartridges from product purification from the production of 1,1-dimethylhydrazine (UDMH) from carboxylic acid hydrazides. K110	....	10	4	K110	X	10 (4.54)
Condensed column overheads from intermediate separation from the production of 1,1-dimethylhydrazine (UDMH) from carboxylic acid hydrazides. K111	....	1*	4	K111	A	10 (4.54)
Product washwaters from the production of dinitrotoluene via nitration of toluene. K112	....	1*	4	K112	A	10 (4.54)
Reaction by-product water from the drying column in the production of toluenediamine via hydrogenation of dinitrotoluene. K113	....	1*	4	K113	A	10 (4.54)

Description			Code		RQ
Condensed liquid light ends from the purification of toluenediamine in the production of toluenediamine via hydrogenation of dinitrotoluene.	1*	4	K114	A	10 (4.54)
Vicinals from the purification of toluenediamine in the production of toluenediamine via hydrogenation of dinitrotoluene.	1*	4	K115	A	10 (4.54)
Heavy ends from the purification of toluenediamine in the production of toluenediamine via hydrogenation of dinitrotoluene.	1*	4	K116	A	10 (4.54)
Organic condensate from the solvent recovery column in the production of toluene diisocyanate via phosgenation of toluenediamine.	1*	4	K117	X	1 (0.454)
Wastewater from the reaction vent gas scrubber in the production of ethylene bromide via bromination of ethene.	1*	4	K118	X	1 (0.454)
Spent absorbent solids from purification of ethylene dibromide in the production of ethylene dibromide.	1*	4	K123	A	10 (4.54)
Process wastewater (including supernates, filtrates, and washwaters) from the production of ethylenebisdithiocarbamic acid and its salts.	1*	4	K124	A	10 (4.54)
Reactor vent scrubber water from the production of ethylenebisdithiocarbamic acid and its salts.	1*	4	K125	A	10 (4.54)
Filtration, evaporation, and centrifugation solids from the production of ethylenebisdithiocarbamic acid and its salts.	1*	4	K126	A	10 (4.54)
Baghouse dust and floor sweepings in milling and packaging operations from the production or formulation of ethylenebisdithiocarbamic acid and its salts.	100	4	K131	X	100 (45.4)
Wastewater from the reactor and spent sulfuric acid from the acid dryer in the production of methyl bromide.	1000	4	K132	X	1000 (454)

| Hazardous substance | CASRN | Regulatory synonyms | Statutory | | | Final RQ | |
			RQ	Code†	RCRA waste Number	Category	Pounds (Kg)
Spent absorbent and wastewater solids from the production of methyl bromide. K136	....	....	1*	4	K136	X	1 (0.454)
Still bottoms from the purification of ethylene dibromide in the production of ethylene dibromide via bromination of ethene. K141	....	....	1*	4	K141	X	1 (0.454)
Process related from the recovery of coal tar, including, but not limited to, tar collecting sump residues from the production of coke by-products produced from coal. This listing does not include K087 (decanter tank tar sludge from coking operations.) K142	....	....	1*	4	K142	X	1 (0.454)
Tar storage tank residues from the production of coke from coal or from the recovery of coke by-products produced from coal. K143	....	....	1*	4	K143	X	1 (0.454)
Process residues from the recovery of light oil, including, but not limited to, those generated in stills, decanters, and wash oil recovery units from the recovery of coke by-products produced from coal. K144	....	....	1*	4	K144	X	1 (0.454)
Wastewater sump residues from light oil refining, including, but not limited to, intercepting or contamination sump sludges from the recovery of coke by-products produced from coal. K145	....	....	1*	4	K145	X	1 (0.454)
Residues from naphthalene collection and recovery operations from the recovery of coke by-products produced from coal. K147	....	....	1*	4	K147	X	1 (0.454)
Tar storage tank residues from coal tar refining. K148	....	....	1*	4	K148	X	1 (0.454)

Residues from coal tar distillation, including, but not limited to, still bottoms.

K149 Distillation bottoms from the production of alpha- (or methyl-) chlorinated toluenes, ring-chlorinated toluenes, benzoyl chlorides, and compounds with mixtures of these functional groups. [This waste does not include still bottoms from the distillation of benzyl chloride.]	1*	4	K149	A	10 (4.54)
K150 Organic residuals, excluding spent carbon adsorbent, from the spent chlorine gas and hydrochloric acid recovery processes associated with the production of alpha- (or methyl-) chlorinated toluenes, ring-chlorinated toluenes, benzoyl chlorides, and compounds with mixtures of these functional groups.	1*	4	K150	A	10 (4.54)
K151 Wastewater treatment sludges, excluding neutralization and biological sludges, generated during the treatment of wastewaters from the production of alpha- (or methyl-) chlorinated toluenes, ring-chlorinated toluenes, benzoyl chlorides, and compounds with mixtures of these functional groups.	1*	4	K151	A	10 (4.54)

† Indicates the statutory source as defined by 1, 2, 3, and 4 below.

†† No reporting of releases of this hazardous substance is required if the diameter of the pieces of the solid metal released is equal to or exceeds 100 micrometers (0.004 inches).

††† The RQ for asbestos is limited to friable forms only.

1—indicates that the statutory source for designation of this hazardous substance under CERCLA is CWA Section 311(b)(4).

2—indicates that the statutory source for designation of this hazardous substance under CERCLA is CWA Section 307(a).

3—indicates that the statutory source for designation of this hazardous substance under CERCLA is CAA Section 112.

4—indicates that the statutory source for designation of this hazardous substance under CERCLA is RCRA Section 3001.

1*—indicates that the 1-pound RQ is a CERCLA statutory RQ.

#BBB indicates that the RQ is subject to change when the assessment of potential carcinogenicity is completed.

#BBB# The Agency may adjust the statutory RQ for this hazardous substance in a future rulemaking; until then the statutory RQ applies.

§—The adjusted RQs for radionuclides may be found in appendix B to this table.

**—indicates that no RQ is being assigned to the generic or broad class.

B

Hazardous Material Labels and Placards

Appendix B presents information about labels and placards that must be posted on most chemicals and hazardous wastes prior to transportation. Labels and placards are placed on shipping containers and transportation vehicles to indicate what is in the load in case of an accident or spill. They also help the handlers in making determinations of the degree of caution to exercise when moving these chemicals and hazardous wastes. Many organizations also use the labels to provide information during use and storage.

Appendix B information is from 40 CFR Chapter 1 and was prepared by the U.S. Department of Transportation.

There are several differences between labels and placards. Size is the biggest visual difference, since labels are only four inches square, whereas placards must be 10 3/4 inches square. Labels are to be used on inidividual containers of chemicals and hazardous wastes with less than 640 cubic feet in volume, such as the common 55 gallon drum. Placards are to be used on containers with a volume greater than 640 cubic feet, such as on transport vehicles.

The labels and placards come in a variety of colors, although they appear here in black and white. A description of colors that must be used on labels and placards is included in the text of Appendix B.

Hazardous Material Labels

(b) In addition to complying with § 172.407, the background color on the EXPLOSIVE 1.1, EXPLOSIVE 1.2 and EXPLOSIVE 1.3 labels must be orange. The "**" shall be replaced with the appropriate division number and compatibility group. The compatibility group letter must be the same size as the division number and must be shown as a capitalized Roman letter.

(c) Except for size and color, the EXPLOSIVE 1.4, EXPLOSIVE 1.5, EXPLOSIVE 1.6 labels, and EXPLOSIVE Subsidiary label must be as follows:

EXPLOSIVE 1.4:

§ 172.411 **EXPLOSIVE 1.1, 1.2, 1.3, 1.4, 1.5 and 1.6 labels, and EXPLOSIVE Subsidiary label.**

(a) Except for size and color, the EXPLOSIVE 1.1, EXPLOSIVE 1.2 and EXPLOSIVE 1.3 labels must be as follows:

EXPLOSIVE 1.5:

EXPLOSIVE Subsidiary label:

EXPLOSIVE 1.6:

(d) In addition to complying with § 172.407, the background color on the EXPLOSIVE 1.4, EXPLOSIVE 1.5, EXPLOSIVE 1.6, and EXPLOSIVE Subsidiary label must be orange. Except for the EXPLOSIVE subsidiary label, the "*" shall be replaced with the appropriate compatibility group. The compatibility group letter must be shown as a capitalized Roman letter measuring at least 12.7 mm (0.5 inches) in height. Except for the EXPLOSIVE subsidiary label, division numerals must measure at least 30 mm (1.2 inches) in height and at least 5 mm (0.2 inches) in width.

[Amdt. 172-123, 56 FR 66256, Dec. 20, 1991]

§ 172.415   NON-FLAMMABLE GAS Label.

(a) Except for size and color, the NON-FLAMMABLE GAS label must be as follows:

(b) In addition to complying with § 172.407, the background color on the NON-FLAMMABLE GAS label must be green.

[Amdt. 172-123, 56 FR 66256, Dec. 20, 1991]

§ 172.416 POISON GAS label.

(a) Except for size and color, the POISON GAS label must be as follows:

(b) In addition to complying with § 172.407, the background color on the FLAMMABLE GAS label must be red.

[Amdt. 172-123, 56 FR 66257, Dec. 20, 1991]

§ 172.419 FLAMMABLE LIQUID label.

(a) Except for size and color the FLAMMABLE LIQUID label must be as follows:

(b) In addition to complying with § 172.407, the background on the POISON GAS label must be white.

[Amdt. 172-123, 56 FR 66257, Dec. 20, 1991]

§ 172.417 FLAMMABLE GAS label.

(a) Except for size and color, the FLAMMABLE GAS label must be as follows:

(b) In addition to complying with § 172.407, the background color on the FLAMMABLE LIQUID label must be red.

[Amdt. 172-123, 56 FR 66257, Dec. 20, 1991]

§ 172.420 FLAMMABLE SOLID label.

(a) Except for size and color, the FLAMMABLE SOLID label must be as follows:

(b) In addition to complying with § 172.407, the background on the FLAMMABLE SOLID label must be white with vertical red stripes equally spaced on each side of a red stripe placed in the center of the label. The red vertical stripes must be spaced so that, visually, they appear equal in width to the white spaces between them. The symbol (flame) and text (when used) must be overprinted. The text "FLAMMABLE SOLID" may be placed in a white rectangle.

[Amdt. 172-123, 56 FR 66257, Dec. 20, 1991]

§ 172.422 SPONTANEOUSLY COMBUS-
TIBLE label.

(a) Except for size and color, the SPONTANEOUSLY COMBUSTIBLE label must be as follows:

(b) In addition to complying with § 172.407, the background color on the lower half of the SPONTANEOUSLY COMBUSTIBLE label must be red and the upper half must be white.

[Amdt. 172-123, 56 FR 66257, Dec. 20, 1991, as amended at 57 FR 45458, Oct. 1, 1992]

§ 172.423 DANGEROUS WHEN WET
label.

(a) Except for size and color, the DANGEROUS WHEN WET label must be as follows:

(b) In addition to complying with § 172.407, the background color on the DANGEROUS WHEN WET label must be blue.

[Amdt. 172-123, 56 FR 66257, Dec. 20, 1991]

§ 172.426 OXIDIZER label.

(a) Except for size and color, the OXIDIZER label must be as follows:

(b) In addition to complying with § 172.407, the background color on the OXIDIZER label must be yellow.

[Amdt. 172-123, 56 FR 66257, Dec. 20, 1991]

§ 172.427 ORGANIC PEROXIDE label.

(a) Except for size and color, the ORGANIC PEROXIDE label must be as follows:

(b) In addition to complying with § 172.407, the background color on the ORGANIC PEROXIDE label must be yellow.

[Amdt. 172-123, 56 FR 66258, Dec. 20, 1991]

§ 172.430 POISON label.

(a) Except for size and color, the POISON label must be as follows:

(b) In addition to complying with § 172.407, the background on the POISON label must be white.

[Amdt. 172-123, 56 FR 66258, Dec. 20, 1991]

§ 172.431 KEEP AWAY FROM FOOD label.

(a) Except for size and color, the KEEP AWAY FROM FOOD label must be as follows:

(b) In addition to complying with § 172.407, the background on the KEEP AWAY FROM FOOD label must be white.

[Amdt. 172-123, 56 FR 66258, Dec. 20, 1991]

§ 172.432 INFECTIOUS SUBSTANCE label.

(a) Except for size and color, the INFECTIOUS SUBSTANCE label must be as follows:

(b) In addition to complying with § 172.407, the background on the INFECTIOUS SUBSTANCE label must be white.

[Amdt. 172–123, 56 FR 66258, Dec. 20, 1991]

§ 172.436 RADIOACTIVE WHITE-I label.

(a) Except for size and color, the RADIOACTIVE WHITE-I label must be as follows:

(b) In addition to complying with § 172.407, the background on the RADIOACTIVE WHITE-I label must be white. The printing and symbol must be black, except for the "I" which must be red.

[Amdt. 172–123, 56 FR 66259, Dec. 20, 1991]

§ 172.438 RADIOACTIVE YELLOW-II label.

(a) Except for size and color, the RADIOACTIVE YELLOW-II must be as follows:

(b) In addition to complying with § 172.407, the background color on the RADIOACTIVE YELLOW-II label must be yellow in the top half and white in the lower half. The printing and symbol must be black, except for the "II" which must be red.

[Amdt. 172–123, 56 FR 66259, Dec. 20, 1991]

§ 172.440 RADIOACTIVE YELLOW-III label.

(a) Except for size and color, the RADIOACTIVE YELLOW-III label must be as follows:

(b) In addition to complying with § 172.407, the background color on the RADIOACTIVE YELLOW-III label must be yellow in the top half and white in the lower half. The printing and symbol must be black, except for the "III" which must be red.

[Amdt. 172-123, 56 FR 66259, Dec. 20, 1991]

§ 172.442   CORROSIVE label.

(a) Except for size and color, the CORROSIVE label must be as follows:

(b) In addition to complying with § 172.407, the background on the CORROSIVE label must be white in the top half and black in the lower half.

[Amdt. 172-123, 56 FR 66259, Dec. 20, 1991]

§ 172.444   [Reserved]

§ 172.446   CLASS 9 label.

(a) Except for size and color, the "CLASS 9" (miscellaneous hazardous materials) label must be as follows:

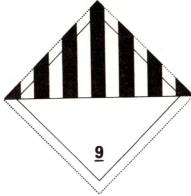

(b) In addition to complying with § 172.407, the background on the CLASS 9 label must be white with seven black vertical stripes on the top half. The black vertical stripes must be spaced, so that, visually, they appear equal in width to the six white spaces between them. The lower half of the label must be white with the class number "9" underlined and centered at the bottom.

[Amdt. 172-123, 56 FR 66259, Dec. 20, 1991]

§ 172.448   CARGO   AIRCRAFT   ONLY label.

(a) Except for size and color, the CARGO AIRCRAFT ONLY label must be as follows:

(b) The CARGO AIRCRAFT ONLY label must be black on an orange background.

[Amdt. 172-123, 56 FR 66259, Dec. 20, 1991]

§ 172.450   EMPTY label.

(a) Each EMPTY label, except for size, must be as follows:

(1) Each side must be at least 6 inches (152 mm.) with each letter at least 1 inch (25.4 mm.) in height.

(2) The label must be white with black printing.

(b) [Reserved]

Hazardous Material Placards

and EXPLOSIVES 1.3 placards must be as follows:

§ 172.521 **DANGEROUS placard.**

(a) Except for size and color, the DANGEROUS placard must be as follows:

(b) In addition to complying with § 172.519 of this subpart, the background color on the EXPLOSIVES 1.1, EXPLOSIVES 1.2, and EXPLOSIVES 1.3 placards must be orange. The "*" shall be replaced with the appropriate division number and, when required, appropriate compatibility group letter. The symbol, text, numerals and inner border must be black.

[Amdt. 172–123, 55 FR 52602, Dec. 21, 1990, as amended at 56 FR 66260, Dec. 20, 1991]

§ 172.523 **EXPLOSIVES 1.4 placard.**

(a) Except for size and color, the EXPLOSIVES 1.4 placard must be as follows:

(b) In addition to meeting the requirements of § 172.519, and appendix B to this part, the DANGEROUS placard must have a red upper and lower triangle. The placard center area and ½-inch (12.7 mm.) border must be white. The inscription must be black with the ⅛-inch (3.2 mm.) border marker in the white area at each end of the inscription red.

[Amdt. 172–29, 41 FR 15996, Apr. 15, 76, as amended by Amdt. 172–29A, 41 FR 40680, Sept. 20, 1976]

§ 172.522 **EXPLOSIVES 1.1, EXPLOSIVES 1.2 and EXPLOSIVES 1.3 placards.**

(a) Except for size and color, the EXPLOSIVES 1.1, EXPLOSIVES 1.2

(b) In addition to complying with § 172.519 of this subpart, the background color on the EXPLOSIVES 1.4 placard must be orange. The "*" shall be replaced, when required, with the appropriate compatibility group letter. The division numeral, 1.4, must measure at least 64 mm (2.5 inches) in height. The text, numerals and inner border must be black.

[Amdt. 172-123, 55 FR 52602, Dec. 21, 1990, as amended at 56 FR 66261, Dec. 20, 1991]

§ 172.524 EXPLOSIVES 1.5 placard.

(a) Except for size and color, the EXPLOSIVES 1.5 placard must be as follows:

(b) In addition to complying with the § 172.519 of this subpart, the background color on EXPLOSIVES 1.5 placard. The "*" shall be replaced, when required, with the appropriate compatibility group letter. The division numeral, 1.5, must measure at least 64 mm (2.5 inches) in height. The text, numerals and inner border must be black.

[Amdt. 172-123, 55 FR 52602, Dec. 21, 1990, as amended ât 56 FR 66261, Dec. 20, 1991]

§ 172.525 EXPLOSIVES 1.6 placard.

(a) Except for size and color the EXPLOSIVES 1.6 placard must be as follows:

(b) In addition to complying with § 172.519 of this subpart, the background color on the EXPLOSIVES 1.6 placard must be orange. The "*" shall be replaced, when required, with the appropriate compatibility group letter. The division numeral, 1.6, must measure at least 64 mm (2.5 inches) in height. The text, numerals and inner bordest be black.

[Amdt. 172-123, 55 FR 52603, Dec. 21, 1990, as amended at 56 FR 66261, Dec. 20, 1991]

§ 172.526 Standard requirements for the RESIDUE placard.

(a) Each RESIDUE placard must be as follows:

(1) Except as provided in paragraph (a)(3) of this section, the lower triangle of the RESIDUE placard must be black and the word "RESIDUE" must be in white letters approximately 25 mm (1 inch) high, made with approximately 6.3 mm (0.25 inch) stroke.

(2) Except for the RADIOACTIVE, EXPLOSIVES 1.1, 1.2, 1.3, 1.4, 1.5 or 1.6, DANGEROUS, or subsidiary placard required by § 172.505 of this subpart, the RESIDUE placard may be used to display the appropriate identification number in accordance with the provisions of subpart D of this part.

(3) For a combustible liquid residue, the lower triangle of the RESIDUE placard must be white and the word "RESIDUE" must be in black letters.

(4) Otherwise, the RESIDUE placard must be as specified in

§§ 172.519, 172.528, 172.530, 172.532, 172.540, 172.542, 172.544, 172.546, 172.547, 172.548, 172.550, 172.552, 172.553, 172.554, 172.558 and 172.560 as appropriate for the residue of the hazardous material being transported and required by this subchapter to be placarded. No other placard may be used as a RESIDUE placard.

(b) The lower part of each placard must be specified in appendix B to this part and as illustrated on the FLAMMABLE-RESIDUE placard which, except for size and color, must be as follows:

(c) The RESIDUE placard must be as shown in paragraph (b) of this section and may be—

(1) A separate placard,

(2) On the reverse side of a placard, or

(3) A composite made according to the specifications in this section. The lower triangle of the appropriate placard should have a black triangle bearing the word RESIDUE in white letters with the appropriate hazard class number in white.

[Amdt. 172-98, 50 FR 39007, Sept. 26, 1985, as amended by Amdt. 172-104, 51 FR 23079, June 25, 1986; Amdt. 172-106, 51 FR 34987, Oct. 1, 1986. Redesignated by Amdt. 172-123, 55 FR 52603, Dec. 21, 1990, and amended at 56 FR 66261, Dec. 20, 1991; 57 FR 45460, Oct. 1, 1992]

§ 172.527 **Background requirements for certain placards.**

(a) Except for size and color, the square background required by § 172.510(a) for certain placards on rail cars, and § 172.507 for placards on motor vehicles containing a package of highway route controlled quantity radioactive materials, must be as follows:

(b) In addition to meeting the requirements of § 172.519 for minimum durability and strength, the square background must consist of a white square measuring 14¼ inches (362.0 mm.) on each side surrounded by a black border extending to 15¼ inches (387.0 mm.) on each side.

(49 U.S.C. 1803, 1804, 1808; 49 CFR 1.53, app. A to part 1)

[Amdt. 172-29, 41 FR 15996, Apr. 15, 1976, as amended by Amdt. 172-64, 46 FR 5316, Jan. 19, 1981; Amdt. 172-78, 48 FR 10226, Mar. 10, 1983]

§ 172.528 **NON-FLAMMABLE GAS placard.**

(a) Except for size and color, the NON-FLAMMABLE GAS placard must be as follows:

(b) In addition to complying with § 172.519, the background color on the NON-FLAMMABLE GAS placard must be green. The letters in both words must be at least 38 mm (1.5 inches) high. The symbol, text, class number and inner border must be white.

[Amdt. 172-123, 56 FR 66261, Dec. 20, 1991]

§ 172.530   OXYGEN placard.

(a) Except for size and color, the OXYGEN placard must be as follows:

(b) In addition to complying with § 172.519 of this subpart, the background color on the OXYGEN placard must be yellow. The symbol, text, class number and inner border must be black.

[Amdt. 172-123, 56 FR 66262, Dec. 20, 1991]

§ 172.532   FLAMMABLE GAS placard.

(a) Except for size and color, the FLAMMABLE GAS placard must be as follows:

(b) In addition to complying with § 172.519, the background color on the FLAMMABLE GAS placard must be red. The symbol, text, class number and inner border must be white.

[Amdt. 172-123, 56 FR 66262, Dec. 20, 1991]

§ 172.536   [Reserved]

§ 172.540   POISON GAS placard.

(a) Except for size and color, the POISON GAS placard must be as follows:

(b) In addition to complying with § 172.519, the background color on the POISON GAS placard must be white. The symbol, text, class number and inner border must be black.

[Amdt. 172-123, 56 FR 66262, Dec. 20, 1991]

§ 172.542 FLAMMABLE placard.

(a) Except for size and color, the FLAMMABLE placard must be as follows:

(b) In addition to complying with § 172.519, the background color on the FLAMMABLE placard must be red. The symbol, text, class number and inner border must be white.

(c) The word "GASOLINE" may be used in place of the word "FLAMMA-BLE" on a placard that is displayed on a cargo tank or a portable tank being used to transport gasoline by highway. The word "GASOLINE" must be shown in white.

[Amdt. 172-123, 56 FR 66262, Dec. 20, 1991]

§ 172.544 COMBUSTIBLE placard.

(a) Except for size and color, the COMBUSTIBLE placard must be as follows:

(b) In addition to complying with § 172.519, the background color on the COMBUSTIBLE placard must be red. The symbol, text, class number and inner border must be white. On a COMBUSTIBLE placard with a white bottom as prescribed by § 172.332(c)(4), the class number must be red or black.

(c) The words "FUEL OIL" may be used in place of the word "COMBUS-TIBLE" on a placard that is displayed on a cargo tank or portable tank being used to transport by highway fuel oil that is not classed as a flammable liquid. The words "FUEL OIL" must be white.

[Amdt. 172-123, 56 FR 66262, Dec. 20, 1991]

§ 172.546 FLAMMABLE SOLID placard.

(a) Except for size and color, the FLAMMABLE SOLID placard must be as follows:

(b) In addition to complying with §172.519, the background on the FLAMMABLE SOLID placard must be white with seven vertical red stripes. The stripes must be equally spaced, with one red stripe placed in the center of the label. Each red stripe and each white space between two red stripes must be 25 mm (1.0 inches) wide. The letters in the word "SOLID" must be at least 38.1 mm (1.5 inches) high. The symbol, text, class number and inner border must be black.

[Amdt. 172-123, 56 FR 66263, Dec. 20, 1991]

§ 172.547 SPONTANEOUSLY COMBUSTIBLE placard.

(a) Except for size and color, the SPONTANEOUSLY COMBUSTIBLE placard must be as follows:

(b) In addition to complying with § 172.519, the background color on the SPONTANEOUSLY COMBUSTIBLE placard must be red in the lower half and white in upper half. The letters in the word "SPONTANEOUSLY" must be at least 25 mm (0.98 inches) high. The symbol, text, class number and inner border must be black.

[Amdt. 172-123, 56 FR 66263, Dec. 20, 1991]

§ 172.548 DANGEROUS WHEN WET placard.

(a) Except for size and color, the DANGEROUS WHEN WET placard must be as follows:

(b) In addition to complying with § 172.519, the background color on the DANGEROUS WHEN WET placard must be blue. The letters in the words "WHEN WET" must be at least 25 mm (1.0 inches) high. The symbol, text, class number and inner border must be white.

[Amdt. 172-123, 56 FR 66263, Dec. 20, 1991]

§ 172.550   OXIDIZER placard.

(a) Except for size and color, the OXIDIZER placard must be as follows:

(b) In addition to complying with § 172.519, the background color on the OXIDIZER placard must be yellow. The symbol, text, division number and inner border must be black.

[Amdt. 172-123, 56 FR 66263, Dec. 20, 1991]

§ 172.552   ORGANIC PEROXIDE placard.

(a) Except for size and color, the ORGANIC PEROXIDE placard must be as follows:

(b) In addition to complying with § 172.519, the background color on the ORGANIC PEROXIDE placard must be yellow. The symbol, text, division number and inner border must be black.

[Amdt. 172-123, 56 FR 66263, Dec. 20, 1991]

§ 172.553   KEEP AWAY FROM FOOD placard.

(a) Except for size and color, the KEEP AWAY FROM FOOD placard must be as follows:

(b) In addition to complying with § 172.519, the background on the KEEP AWAY FROM FOOD placard must be white. The size of the lettering below the word "HARMFUL" must be proportional to that shown.

The symbol, text, class number and inner border must be black.

[Amdt. 172-123, 56 FR 66263, Dec. 20, 1991]

§ 172.554 POISON placard.

(a) Except for size and color, the POISON placard must be as follows:

(b) In addition to complying with § 172.519, the background on the POISON placard must be white. The symbol, text, class number and inner border must be black.

[Amdt. 172-123, 56 FR 66264, Dec. 20, 1991]

§ 172.556 RADIOACTIVE placard.

(a) Except for size and color, the RADIOACTIVE placard must be as follows:

(b) In addition to complying with § 172.519, the background color on the RADIOACTIVE placard must be white in the lower portion with a yellow triangle in the upper portion. The base of the yellow triangle must be 29 mm ± 5mm (1.1 inches ±z0.2 inches) above the placard horizontal center line. The symbol, text, class number and inner border must be black.

[Amdt. 172-123, 56 FR 66264, Dec. 20, 1991]

§ 172.558 CORROSIVE placard.

(a) Except for size and color, the CORROSIVE placard must be as follows:

(b) In addition to complying with § 172.519, the background color on the CORROSIVE placard must be black in the lower portion with a white triangle in the upper portion. The base of the white triangle must be 38 mm ± 5mm (1.5 inches ± 0.2 inches) above the placard horizontal center line. The text and class number must be white. The symbol and inner border must be black.

[Amdt. 172-123, 56 FR 66264, Dec. 20, 1991]

§ 172.560 CLASS 9 placard.

(a) Except for size and color the CLASS 9 (miscellaneous hazardous materials) placard must be as follows:

(b) In addition to conformance with § 172.519, the background on the CLASS 9 placard must be white with seven black vertical stripes on the top half extending from the top of the placard to one inch above the horizontal centerline. The black vertical stripes must be spaced so that, visually, they appear equal in width to the six white spaces between them. The space below the vertical lines must be white with the class number 9 underlined and centered at the bottom.

[Amdt. 172-123, 56 FR 66264, Dec. 20, 1991, as amended at 57 FR 45460, Oct. 1, 1992]

FREQUENTLY USED ACRONYMS AND ABBREVIATIONS

ACGIH	American Conference of Governmental Industrial Hygienists
ANSI	American National Standards Institute
BACT	Best available control technology
BAT	Best available technology
BDAT	Best demonstrated available technology
BMP	Best management practice
BPCT	Best practical control technology
BS 7750	British Standard 7750—Environmental Management Systems
C	Temperature in degrees Celsius
CAS	Chemical abstract service
CAA	Clean Air Act
CCR	California Code of Regulations
CERCLA	Comprehensive Environmental Response, Compensation and Reauthorization Act
CFR	Code of Federal Regulations
CHP	California Highway Patrol
CWA	Clean Water Act
DOE	U.S. Department of Energy
DOT	U.S. Department of Transportation
EIS	Environmental impact statement
EPA	U.S. Environmental Protection Agency
ER	Emergency response

FR	Federal Register
HMBP	Hazardous material business plan
HMMP	Hazardous material management plan
HMTA	Hazardous Material Transportation Act
HMW	Hazardous material (chemical) and hazardous waste
HW	Hazardous waste
HWCL	Hazardous Waste Control Law
ISO	International Standards Organization
LC 50	Concentration lethal to 50% of sample
LD	Lethal dose
LDR	Land disposal restriction
LQG	Large-quantity generator
MSDS	Material safety data sheet
NEPA	National Environmental Protection Act
NIOSH	National Institute for Occupational Safety and Health
NFPA	National Fire Protection Association
NOAA	National Oceanographic and Atmospheric Agency
NPDES	National pollutant discharge elimination system
NPL	National Priority List
OSHA	U.S. Occupational Safety and Health Administration
PEL	Permissible exposure limit
pH	Measure of acidity or alkalinity
POTW	Publicly owned treatment works
ppb	Parts per billion
PPE	Personal protective equipment
ppm	Parts per million
PRP	Potentially responsible party

RCRA	Resource Conservation and Recovery Act
RMPP	Risk management prevention plan
RQ	Reportable quantity
RWQCB	California Regional Water Quality Control Board
SARA	Superfund Amendments and Reauthorization Act
SB	California Senate Bill
SCBA	Self-contained breathing apparatus
SPCCP	Spill prevention control and countermeasure plan
SQG	Small-quantity generator
STLC	Soluble threshold limit concentration
TCLP	Toxicity characteristic leaching procedure
TLV	Threshold limit value
TSCA	Toxic Substances Control Act
TSDF	Treatment, storage and disposal facility
TTLC	Total threshold limit concentration
TTU	Transportable treatment unit
TWA	Time-weighted average
UFC	Uniform Fire Code
UST	Underground storage tank

GLOSSARY

(These are "Working Definitions," Not Strict Regulatory Definitions)

Abatement   Clean-up actions that reduce the amount of pollution.

Abrasive Blasting   A clean-up method that uses dry or wet abrasive particles under air pressure to remove contamination from metal, concrete and other surfaces.

ACGIH   The American Conference of Governmental Industrial Hygienists, which recommends safe exposure levels for various substances. For example, numerous chemicals have been assigned threshold limit values by the ACGIH.

Acid Cleaning   The use of acids for the purpose of removing contaminants, usually from surfaces.

Activated Carbon   A granular or powdered substance used to remove many organic and inorganic substances from liquids via adsorption. The contaminant is incorporated onto a carbon matrix.

Acute Exposure   A high exposure for a short duration.

Asbestos   Mineral fibers ranging in length from 0.1 to 10 micron and are in friable or non-friable form. Asbestos can cause asbestosis, lung cancer and mesothelioma.

Atomic Absorption Spectrometry   A laboratory method in which the sample is atomized and a light source is introduced. Characteristic wavelengths of light are absorbed by certain atoms allowing identification. The quantity of the substance present is determined by the light's intensity.

Biennial Reports   A report required by some agencies, which presents wastes generated and minimized. The source of the data is manifests.

Bill of Lading   A document required for the transportation of hazardous materials. The bill must accompany the load.

Bioaccumulation   The process by which a pollutant accumulates in the tissues of an organism.

Biodegradation   Natural or introduced microorganisms, such as bacteria, utilize certain organic contaminants as a source of food in their metabolic process.

Carcinogen   An agent which causes cancer or an abnormal growth of cells.

Centrifugation   A physical treatment process that utilizes centrifugal force to concentrate contaminants with different densities.

Chemical Neutralization   The simple neutralization of an acid by adding a base or the reverse.

Chemical Oxidation   Used for the treatment of cyanide and other wastes. Oxidation of the cyanide occurs by using chlorine, ozonation, hydrogen peroxide or electro-chemical oxidation.

Chemical Reduction   The addition of strong reducing agents, such as sulfur dioxide and sodium bisulfite.

Chronic Exposure   A lower exposure for a longer duration.

Code of Federal Regulations   A collection of final Federal laws and regulations arranged by subject and updated one time per year.

Contingency Plan   Planning that allows an organization to be prepared for an emergency before it occurs. For example, documents should specify what should happen in case of a chemical spill, fire or other emergency.

Corrosive Materials   A liquid or solid that can corrode metals or destroy human tissue by skin contact or inhalation.

Deep Well Injection   This system is used for limited disposal of some liquid hazardous wastes. The waste cannot be injected into or above potable water supplies.

Dose Effect   The response to a chemical insult. Usually, as the dose increases, the effect will become progressively more obvious.

Disposal   The placement of hazardous waste in various environments, including land, air, injection wells, oceans, and so forth. In most locations, laws require that the hazardous waste has been treated to specified levels prior to disposal.

Electrical Resistivity   The assessment of subsurface electrical resistivities in ground water, rock and soil. The system depends on the application of electrical currents into the ground via surface electrodes.

EPA Identification Number   A number which is assigned by the EPA prior to shipment of hazardous waste off-site. The number is also required for transporters and treatment, storage and disposal facilities (TSDFs).

Fixation/Solidification   The reduction of mobility of a hazardous chemical by fixing the toxin into a non-leachable matrix.

Flash Point   The lowest temperature at which the vapors of a liquid decompose and become a flammable gaseous mixture.

Flotation   Suspended solids are caused to float to the surface of a tank, where they can be removed.

Gas Chromatography   The separation of a variety of organic compounds and identification using several different types of detectors.

Generator   Any person whose act first causes a hazardous waste to become subject to regulation. An organization usually becomes a generator when it spills, uses or no longer plans to use a hazardous material.

Ground Penetrating Radar   Subsurface materials with various electrical properties will reflect differently. In this system, high-frequency radio waves are introduced into the ground. A receiving antenna then picks up the reflected return wave.

Hazardous Material   Chemicals that are capable of posing a risk to health, safety or property.

Hazardous Material Business Plan   A document containing hazardous material management information, site-specific maps and chemical inventories.

Hazardous Material Management Plan   A document that pertains primarily to underground storage tanks and presents information about types of tanks, leak detection and material stored.

Hazardous Waste   Waste substances that by nature are inherently dangerous to handle or to dispose of, may cause death or illness or be a hazard to health or the environment when improperly handled. These are usually hazardous materials (chemicals) that have been used, spilled or are no longer needed.

Hazard Recognition/Communication Plan   Plans pertaining to training for field forces in safety matters, such as recognition and safe handling of hazardous materials. The plan should be written by all organizations with over 10 employees to meet OSHA requirements.

Ignitibility   Substances that meet one of the following criteria: has a flash point less than 60° C, capable of causing fire through friction or absorption of moisture, is a flammable compressed gas or an oxidizer.

Incineration   A detoxification type of treatment utilizing oxidation in high-temperature incinerators. Infectious waste is commonly incinerated.

Inductively Coupled Plasma Emission Spectrometry   The sample is nebulized into a plasma with an input of energy from an argon torch. When the excited electrons relax and return to their normal energy state, they emit a characteristic wavelength of light.

Ion Exchange   Ions are exchanged for ions of similar charge from the solution in which the resin is immersed. The entire exchange process occurs on the surface of the resin.

Landfill   Specially designed locations for placement of primarily solid waste. These include Class I (hazardous waste), Class II (certain large-volume, lower-toxicity hazardous waste) and Class III (garbage).

LD50   The amount of material that will kill 50% of test animals, expressed in mg/kg of body weight.

Leachate   Contaminants that drain out of solid or liquid hazardous waste.

Liner   Membranes that are placed under or on the sides of impoundments, hazardous waste disposal areas and certain chemical storage areas to prevent soil and ground water contamination.

Magnetometry   A proton precession magnetometer is used to measure magnetic forces. Once the earth's normal magnetic field is accounted for, presence of buried ferrous metals (tanks, drums, and so on) can be inferred using this equipment.

Manifest   A hazardous waste manifest is required of any generator of a hazardous waste who transports the waste (or has it transported) or turns it over for treatment, storage or disposal. It is also used as a means of making sure that all hazardous waste is tracked from the point of generation to the point of disposal.

Material Safety Data Sheet   A source of information about products, oriented to protect employees and the community who might be exposed during handling, storage and disposal.

Monitoring Well   A well that is used to assess ground water pollution, quality, levels, and aquifer characteristics.

Mutagen   An agent that changes the character of a gene in a cell; this change is perpetuated in subsequent divisions of that cell. A mutagen alters DNA.

National Priorities List   A list of Superfund sites identified by the EPA.

National Pollutant Discharge Elimination System (NPDES) An NPDES permit is required before discharges can be made to the waters (i.e., streams, rivers, lakes, and so forth) of the United States.

Organic Compounds Compounds that contain carbon and one or more other element.

Organic Vapor Analyzers Equipment commonly used in the field to test for volatile organic compounds. Most of the detectors are based on the ionization potential of the organic compound. Flame and photoionization analyzers are two examples of this type of instrument.

Oxidizers Substances that contain significant amounts of chemically-bonded oxygen. They yield oxygen readily and therefore stimulate combustion of other substances.

Permissible Exposure Levels (PELs) PELs are published safe levels of various substances. PELs are calculated as time-weighted-averages, and, in many cases, they equal a Threshold Limit Value.

pH A measure of the hydrogen ion concentration ranging from 0 (most acidic) to 14 (most basic or alkaline). A value of 7 is neutral.

Polychlorinated Biphenyls (PCBs) PCBs are a group of manufactured chemicals including about 70 different compounds made up of carbon, hydrogen and chlorine. If released to the environment, they persist for long periods of time and can biomagnify in food chains.

Portable X-Ray This type of x-ray fluorescence system is used to identify the presence of certain metals such as arsenic, lead or mercury.

Pretreatment On-site treatment usually done to reduce volume or toxicity prior to discharge into a public-owned treatment works.

Public-Owned Treatment Works (POTW) A sewage treatment facility, usually operated by the city, which in some cases has primary, secondary and tertiary treatment systems. A POTW normally cannot treat hazardous waste.

Pyrophoric Capable of igniting spontaneously when exposed to air.

Radioactive Materials Substances that release alpha, beta and/or gamma radiation and are defined by the Department of Transportation as material having a specific activity greater than $0.002 \, \mu \, Ci/g$.

Representative Sample A sample that is collected in a way that increases its chance of accurately representing true contamination or environment.

Respirable Particles   Particulates and fibers that, when inhaled by humans, are usually retained in the lungs. They range in size from 0.2 to 10 μ.

Reverse Osmosis   A membrane system that can be used to concentrate ions on one side of a membrane and solute on the other. High pressure is used to get the ions to move to the side with the higher ionic concentration (the reverse of osmosis).

Risk Management Prevention Program (RMPP)   A plan required by some regulators if a location stores or uses a significant quantity of extremely or acutely hazardous material. The emphasis in the RMPP is on emergency response planning.

Sedimentation   Gravitational settling that removes some particles from a liquid waste. Commonly, chemicals are added to cause flocculation or coagulation, which produces faster-settling particles.

Small-Quantity Generator   A generator that produces less than 100 kg of hazardous waste per month or less than 1 kg of acutely hazardous waste per month.

Solvent Extraction   Used for the treatment of organic wastes. It helps in volume reduction since the organics are separated from water or nonhazardous solids.

Superfund   The Comprehensive Environmental Response, Compensation and Liability Act.

Teratogen   An agent that disturbs the growth process of an embryo causing a malformed fetus to be formed. A teratogenic effect is a one-generation effect and will not occur in subsequent generations.

Threshold   The lowest concentration of a chemical that produces an observable effect.

Time-Weighted Averages   A way of expressing employee exposure to a hazardous material or waste, based on 480 minutes.

Total Suspended Solids (TSS)   Particles of all sizes that are suspended in a measured volume of water. TSS reduces light penetration in the water column, can clog the gills of fish and are often associated with toxic pollutants because organic materials and metals tend to bind to particles.

Toxic Substance   A chemical or mixture that may present an unreasonable risk of injury to health or the environment.

Toxicity   The amount or ability of a substance to produce an adverse effect. The amount of health effect at a certain level.

Toxicity Characteristic Leaching Procedure   The EPA test specifically designed to determine the mobility of both organic and inorganic contaminants in landfills. The test replaced the EP Toxicity Test on September 25, 1990.

Toxic Waste   A waste that is capable of killing, injuring or damaging an organism. Toxic waste is a type of hazardous waste.

Transporter   The carrier of a hazardous material or waste. The carrier can also become a generator if its vehicle spills, crosses U.S. boundaries or accumulates sludges.

Treatment   Processes designed to reduce the volume and/or toxicity of contaminants. Treatment can be divided into concentration, detoxification and fixation types.

Underground Storage Tank   Any tank with at least 10% of its volume buried below the ground, including any attached pipes.

Vitrification   Melting or fusing of wastes into glass or glassy substances in order to reduce the mobility of the toxic component.

Volatile   Substances that evaporate at lower temperatures.

References

ADAMS, J., "France Bans German Waste," *Waste Tech News,* Sept. 92: 12.

ALLEGRI, T. H., *Handling and Management of Hazardous Materials and Waste.* New York: Chapman and Hall, 1986.

"All packed up and no place to go," *Time*, May 6, 1991, P. 27.

BAERT, G.R., "Hazardous Waste Storage and Shipment Containers." Unpublished Paper, 1990.

BRICKNELL, K., "Historical Rise of Hazardous Material and Waste Management." Unpublished Paper, 1990.

"Building Blocks of an Environmental Compliance Program." San Francisco: Environmental Law Foundation, 1993.

California Environmental Protection Agency, *Hazardous Waste Related Fees.* Sacramento: Cal EPA, 1993.

CROUTH, G., *How to Recognize a Hazardous Waste.* Pittsburgh: Digby Books, 1989.

DENTON, W. J., ET AL., *Environmental Due Diligence Handbook.* Rockville, MD: Government Institute Inc., 1989.

DUNLAP, W. J,. ET AL., "Sampling for Organic Chemicals and Microorganisms in the Subsurface." Robert S. Kerr Environmental Research Laboratory, EPA-600/2-77-176, 1977.

ELLIOTT, J.F. AND D. MORELL, "Hazardous Materials Program Commentary." Vancouver: STP Specialty Technical Publishers, 1990.

ENVIRONMENTAL PROTECTION AGENCY, *Tracking Waste from Cradle to Grave.* Washington: Government Printing Office, 1993.

———. *EPA Hazard Classes.* 1993.

———. *Superfund Enforcement Speaker's Kit.* 1993.

———. *List of Hazardous Substances and Reportable Quantities.* 40 CFR Ch. 1. 1993.

————. *Schematic of Representative Hazardous Waste Landfill.* 1993.

————. *Hazardous Waste Management: What's the Best Approach?* 1993.

————. *Alternative Treatment Technologies Selected for NPL Sites Through Fiscal Year 1991.* EPA/TIO. 1993.

————. *Standard Operating Safety Guides.* 9285.1-03. 1992.

————. *Payment of Superfund Costs.* CERCLIS. 1992.

————. *Hazardous Waste Data Management System.* 1991.

————. *Handbook on In Situ Treatment of Hazardous Waste-Contaminated Soils.* EPA/540/2-90/002. 1990.

————. *Does Your Business Produce Hazardous Waste?* EPA/530-90-027. 1990

————. *RCRA Orientation Manual.* 1990

————. *Definition of a Hazardous Waste and Toxicity Characteristic Leachate Procedure.* 40 CFR. 1990

————. *Requirements for Hazardous Waste Landfill Design, Construction and Closure.* EPA/625/4-89/002. 1989.

————. *Determining Soil Response Action Levels Based on Potential Contaminant Migration to Groundwater: A Compendium of Examples.* PM90-183575. 1989.

————. "Environmental Data: Providing Answers or Raising Questions." *EPA Journal,* 15/3. 1989.

————. *Guidance on Remedial Actions for Contaminated Groundwater at Superfund Sites.* EPA/540/G-88/003. 1988.

————. *The Waste System.* 1988.

————. "The Disposable Society." *EPA Journal,* 14/7. 1988.

————. Compendium of Costs of Remedial Technology at Hazardous Waste Sites. EPA/600/2-87/087. 1987.

————. "Protecting the Earth: Are Our Institutions Up to It?" *EPA Journal,* 15/4. 1987.

————. "Collection of Hazardous Waste Samples." *EPA Journal,* 13/9. 1987.

————. *Solving the Hazardous Waste Problem/EPA's RCRA program.* EPA 530-SW-86-037. 1986.

————. *Mobile Treatment Technologies for Superfund Wastes.* 540/2-86/003(f). 1986.

————. *Superfund Remedial Design and Remedial Action Guidance.* PB8 8-107529. 1986.

————. *Solving the Hazardous Waste Problem/EPA's RCRA Program.* 530-SW-86-037. 1986.

————. *CERCLA/Superfund Orientation Manual.* No date.

————. *Training Manual for Hazardous Waste Site Investigations.* No Date.

"Freight Train Derails, Forces Wide Evacuations," *San Jose Mercury News,* June 30, 1992.

FRIEDMAN, F.B., *Practical Guide to Environmental Management.* Washington: Environmental Law Institute, 1988.

GUBBER, S., "Laboratory Analysis of Hazardous Materials and Wastes." Unpublished Paper, 1990.

HABER, M., "Hazardous Waste Storage Permit and EPA ID Number Requirements—Federal Law." Unpublished Paper, 1993.

HAZTECH SYSTEMS INC., *Solids HAZCAT and Liquids HAZCAT.* Mountain View, CA: Haztech, 1993.

HCL LABELS, INC., *DOT/HM181 Hazardous Materials Label and Placarding Charts.* Sunnyvale, CA: HCL Labels, Inc., 1993.

HIGGINS, T., *Hazardous Waste Minimization Handbook.* MI: Lewis Publishers, 1989.

"Justices Void Ruling on Incinerator Ash." *San Francisco Chronicle,* November 17, 1992.

KEHOE, P., "Major Ecosystems Impacted by Hazardous Materials and Wastes." Unpublished Paper, 1990.

KELLER, J., *Straight Bill of Lading.* WI: J.J. Keller Company, 1993.

LABELMASTER, *Hazardous Waste Label.* Chicago: Labelmaster, 1993.

MACK, J.P. AND T.J. MORAHAN, *Equipment for Data Collection at Hazardous Waste Sites— An Overview for Environmental Professionals.* Sampling and Monitoring Monograph Series, Vol. 1. Silver Springs, MD: HNRGI.

MEEHAN, P., Unpublished Paper prepared for the University of San Francisco. San Francisco: 1993.

"No free rides for Disney," *Los Angeles Times,* 109: 26, July 21, 1990.

"Nuclear Garbage," *Garbage Magazine,* Part 2, May, 1992.

"Paper Firm Fined $13 Million for Pollution," *San Francisco Chronicle,* 1992.

PENNZOIL COMPANY, *Material Safety Data Sheet for Anti-freeze and Summer Coolant.* Pennsylvania: Pennzoil Co., 1993.

ROMIC CHEMICAL CORPORATION, *Illustrations of Unacceptable Conditions.* East Palo Alto, CA: Romic Chemical Corporation, 1993.

SCHOEPKE, R. A. AND K.O. THOMSEN, "Use of Resistivity Sounding to Define the Subsurface Hydrogeology in Glacial Sediments," Proceedings of the 5th National Outdoor Action Conference, 1991.

Second Circuit Court., *Wagner Seed Co. v. Daggett.* 800F.

SIMONSEN, C., *Environmental Law Resource Guide.* Deerfield, IL: Clark Boardman Callaghan, 1994.

Uniform Hazardous Waste Manifest, Little Rock: Arkansas State, 1984.

WEDIN. R., "Pollution Prevention: When Less is Better, Today's Chemist at Work," August 1992, p. 19-22.

WENTZ, C.A., *Hazardous Waste Management.* New York: McGraw Hill, 1989.

WERT, K., "Costs Involved with Hazardous Material and Waste Management." Unpublished Paper, 1990.

INDEX